Pathways
for Inspired
Scientists and
Engineers

Tilahun M. Getu

Pathways for Inspired Scientists and Engineers

Volume I – Pitfalls and Development Opportunities of a Ph.D.

Springer

Tilahun M. Getu
École de Technologie Supérieure (ÉTS)
Montreal, QC, Canada

ISBN 978-3-032-06783-8 ISBN 978-3-032-06784-5 (eBook)
https://doi.org/10.1007/978-3-032-06784-5

This Springer imprint is published by the registered company Springer Nature Switzerland AG
The registered company address is: Gewerbestrasse 11, 6330 Cham, Switzerland

To Atsede Admassu and Melkamu Getu, my parents, for their nurture and sacrifices.

Foreword

I met Dr. Tilahun M. Getu many years ago in Maui, Hawaii, at a scientific conference, when he was at the beginning of his academic journey. We had wonderful and inspirational discussions about science, engineering, and academia. Luckily we stayed connected and exchanged ideas about this topic throughout the years. I am very happy to see the results of his reflections in this book.

It is with deep enthusiasm and admiration that I introduce this vital contribution to the next generation of scientists, engineers, and innovators. *Pathways for Inspired Scientists and Engineers: Volume I – Pitfalls and Development Opportunities of a Ph.D.* and *Pathways for Inspired Scientists and Engineers: Volume II – Skills Acquisition and Career Opportunities of a Ph.D.* arrive at a time when curiosity, resilience, and vision are more essential than ever. This two-volume book is a timely companion for every young mind venturing into the world of research—a world filled not only with opportunities for discovery but also with moments of challenge and growth.

The Ph.D. journey is far more than the pursuit of new knowledge; it is an act of self-transformation. Along this path, each setback becomes a teacher, and every obstacle is an invitation to deepen your understanding—of your field, your colleagues, and, most importantly, yourself. Dr. Getu's work is a testament to this idea. Drawing on years of personal research, mentoring, and his unique perspective as a founder and advocate for young scholars, he weaves a guide that is practical, honest, and truly inspiring.

If you are holding this book, you are already taking bold steps toward a future of meaning and impact. This is not just a roadmap through the technicalities of a Ph.D.—it is an encouragement to embrace the uncertainties, to see pitfalls as stepping stones, and to cultivate the skills, habits, and mindset that will serve you for a lifetime. Science is not just about finding answers; it is about learning to ask the right questions, to persist when the answers seem elusive, and to contribute with integrity and imagination to the broader human story.

To every young scientist and engineer: Let this book be a reminder that you are part of a living tradition of inquiry and hope. The world needs your passion, your ideas, and your courage. May you find in these pages the wisdom to navigate the inevitable storms and the inspiration to shape a brighter, more just, and more innovative tomorrow.

With every good wish on your journey!

Vienna, Austria
July 2025

Schahram Dustdar

Preface

Motivation

A number of authors published excellent books on how to get a Ph.D., mainly reflecting on the overall process of achieving a Ph.D. and managing a doctoral study. Despite the important optimism and wisdom offered by these books, many Ph.D. graduates recall an arduous Ph.D. journey filled with unpleasant circumstances and uncertainties. Although a Ph.D. is supposed to be a challenging intellectual journey that affords various personal and professional opportunities, many graduate students fail to earn or decide not to pursue a Ph.D. due to the global research enterprise's inevitable uncertainties, which translate to numerous specific and generic Ph.D. pitfalls inevitable during the initial, middle, and final years of a Ph.D. study. However, no book has yet exposed the many inevitable Ph.D. pitfalls; offered detailed arguments, perspectives, and strategies on overcoming the inescapable Ph.D. pitfalls; presented the several professional and personal opportunities of a Ph.D.; and discussed the often uncertain transition from a Ph.D. to a bright future.

Driven by the twenty-first century's societal and technological grand challenges, interdisciplinary, multidisciplinary, and transdisciplinary research innovations are dynamically shaping the twenty-first-century research frontiers that inspire the contemporary convergence of science, technology, engineering, and mathematics (STEM). Should one aspire to conduct novel scholarship of discovery, integration, application, and teaching, STEM convergence requires developing depth and breadth of expertise, which would be a significant price to pay for tomorrow's convergence researchers. To rise to this challenge, new-generation researchers should have enthusiasm, purpose, and inspiration in STEM while being intrinsically motivated. To incubate such researchers, there shall be a book that inspires and empowers future-generation researchers toward a Ph.D. and beyond. This need could not be more paramount for the twenty-first-century academia, whose changing forces are making (co-)principal investigators exceedingly busy to rigorously develop new-generation scholars of passion, principle, and purpose.

Scope and Significance

This book originated from the author's years of volunteering at the Institute of Young Ethiopian Scholars! (I-YES!), which was a not-for-profit he had served as a founder, volunteer, and coordinator from August 2013 to August 2018 while shouldering the I-YES! mission of "Inspiring and empowering the Ethiopian and global youth via priceless information and original out-of-the-box philosophies." Informed by 10 years (2013–2023) of research on inspiring and empowering minds and six years (2019–2025) of multidisciplinary neuroscience research, this book is a condensation of a gigantic body of research conducted by the author who has 14 years (2011–2025) of graduate, doctoral, and post-doctoral research experiences in North America and Europe. This book considers a successful Ph.D. journey as *a motivated and thoughtful sailing amid pitfalls*. Like a sailor who arrives at one's destination by understanding the challenges of sailing in bad weather, a Ph.D. student can reach the final moments of a Ph.D. by understanding and overcoming the Ph.D. pitfalls, which are opportunities for skills acquisition, professional development, and personal development.

Drawing on the realms of research philosophy, philosophy of science, psychology of science, motivation science, critical thinking, creative thinking, imaginative thinking, strategic thinking, probabilistic thinking, causal thinking, and independent thinking, this two-volume book offers numerous perspectives, arguments, ideas, and strategies on the pitfalls, professional and personal development opportunities, skills acquisition, and career opportunities of a Ph.D. As a pragmatic guide, this book comprises two volumes: *Volume I – Pitfalls and Development Opportunities of a Ph.D.* and *Volume II – Skills Acquisition and Career Opportunities of a Ph.D.* These volumes are designed for current and prospective Ph.D. students who aspire to conduct original research across all fields of STEM and STEM convergence.

Organization and Target Audience

As in Table 1, nine chapters comprise Volume I, which is aimed at current and prospective master's students in their final year, bachelor's honors students transitioning directly into doctoral programs, master's-plus-Ph.D. students, professionals seeking to accelerate their careers via a professional Ph.D., and industry professionals intending to speed up their careers through an industrial Ph.D. As per Table 2, nine chapters also comprise Volume II, which is intended for all Ph.D. students enrolled at universities all over the globe.

Anyone interested in research philosophy, free thinking, imaginative thinking, critical thinking, creative thinking, strategic thinking, probabilistic thinking, causal thinking, and independent thinking may also find this book important.

Table 1 Organization of *Volume I—Pitfalls and Development Opportunities of a Ph.D.*

Parts	Chapters
Part I: Setting the Stage for a Ph.D.	Chapter 1: What Is a Ph.D.?
	Chapter 2: Why Are Pitfalls Inevitable in a Ph.D.?
Part II: Pitfalls of a Ph.D.	Chapter 3: Research-Related Pitfalls of a Ph.D.
	Chapter 4: Writing and Publication-Related Pitfalls of a Ph.D.
	Chapter 5: Management-Related Pitfalls of a Ph.D.
	Chapter 6: Persona and Motivation-Related Pitfalls of a Ph.D.
	Chapter 7: Generic Pitfalls of a Ph.D.
Part III: Development Opportunities of a Ph.D.	Chapter 8: Professional Development Opportunities of a Ph.D.
	Chapter 9: Personal Development Opportunities of a Ph.D.

Table 2 Organization of *Volume II—Skills Acquisition and Career Opportunities of a Ph.D.*

Parts	Chapters
Part IV: Skills Acquisition of a Ph.D.	Chapter 10: Skills Acquisition in Research
	Chapter 11: Skills Acquisition in Writing and Publication
	Chapter 12: Skills Acquisition in Management
	Chapter 13: Skills Acquisition in Motivation and Rigorous Persona
	Chapter 14: Skills Acquisition in Thinking
	Chapter 15: Skills Acquisition in Imagination
Part V: Career Opportunities of a Ph.D.	Chapter 16: Conventional Careers for a Bright Future
	Chapter 17: Unconventional Careers for a Bright Future
	Chapter 18: Hybrid Careers for a Bright Future

Higher Purpose

Via this *two-volume book of inspiration, imagination, and thinking*, we aim to inspire and empower tomorrow's scientists, technologists, engineers, and mathematicians so they can achieve great research careers beyond their prospective Ph.D. degrees. While aiming to *democratize research*, we particularly intend to inspire the forthcoming-generation researchers to be not only voracious readers but also intensive thinkers and doers who collectively and courageously attack the twenty-first century's scientific, societal, and technological grand challenges. This is the higher purpose—i.e., *telos*—of this book.

Godspeed!

Aspen Hill, MD, USA
July 31, 2025

Tilahun M. Getu

Acknowledgments

For their nurture and sacrifices, Dr. Tilahun M. Getu is indebted to his parents, to whom this book is dedicated. He is also indebted to his wife Kongit, son Joy, and daughter Zoey for making his scientific and non-scientific life vibrant. He acknowledges his family members—Hana, Muluwork, Desalegn, Dawit, Solomon, Kassahun, Beyene, Dr. Barten, and Dr. Rau—for their encouragement and support.

Dr. Getu acknowledges his two anonymous book reviewers for their profound insight, objective criticisms, and invaluable recommendations. He also acknowledges his Springer editor, Ralf Gerstner, for valuable discussions, insightful leadership, and support. He also acknowledges Nargees Subath and her team at Straive, a service provider of Springer Nature, for copy-editing service. He also acknowledges his Ph.D. advisors—Prof. Wessam Ajib (University of Quebec at Montreal (UQAM), QC, Canada) and Prof. René Jr. Landry (École de Technologie Supérieure (ÉTS), QC, Canada)—for their thoughtful leadership, support, and immense research freedom. He also acknowledges his post-doctoral advisor, Prof. Georges Kaddoum (ÉTS, QC, Canada), for his thoughtful leadership, encouragement, and support. He also acknowledges his post-doctoral research collaborators—Prof. Mehdi Bennis (University of Oulu, Finland), Prof. Walid Saad (Virginia Tech, VA, USA), Prof. Schahram Dustdar (Vienna University of Technology, Austria), and Prof. Tadilo E. Bogale (North Carolina A&T State University, NC, USA)—for their thoughtful leadership and support.

Dr. Getu acknowledges the support and encouragement of seasoned scientific leaders such as Prof. Georgios B. Giannakis (University of Minnesota, MN, USA), Prof. Schahram Dustdar (Vienna University of Technology, Austria), Prof. Franz Faupel (University of Kiel, Germany), Prof. Axel Bangert (University of Kassel, Germany), and Prof. Dirk Dahlhaus (University of Kassel, Germany). He also acknowledges his former UQAM colleagues—Dr. Tri Nguyen (Tata Communications, NJ, USA), Prof. Animesh Yadav (Ohio University, OH, USA), and Dr. Mathew Goonewardena (Ericsson, QC, Canada)—for numerous discussions on science, technology, engineering, mathematics, and society. He also acknowledges the many researchers who have enriched his research perspectives through personal communications and interactions at seminars, workshops, and conferences. Last but not least, he acknowledges the research communities of *IEEE*, *ACM*, *AAAI*, *APS*, *AAAS*, *IMS*, *AMS*, *INFORMS*, *SfN*, and *APA* for their collective wisdom and scientific contributions, from which he has continued to learn.

Competing Interests The author declares he has no competing interests that are relevant to the content of this manuscript.

Disclaimer

The realities of your research field and its progress, your research commitment, your doctoral research management, and your life as a Ph.D. student—as perceived by your mind—are all the determining factors in your quest for success in a doctoral study. Because there are *infinite probabilities* between 0 (the probability of an improbable occurrence) and 1 (the probability of a highly likely happening), predicting the likelihood of every dynamic in a Ph.D. student's life and research is hardly possible. Accordingly, despite our years of endeavor to realistically and thoughtfully reflect on the dynamics of Ph.D. students' lives and research, this two-volume book may not necessarily touch each difficulty and reality of every Ph.D. student. Since your research director would have a much better estimate of your realities,[1] difficulties, liking/disliking, and academic background, along with their expectations and research project, the *ultimate guide to your Ph.D. is your Ph.D. (co-)supervisor.*

Therefore, we provide this book "as is" without any warranty or liability. At last, we affirm that the perspectives presented in this book are our own and do not necessarily represent the views of our current or past employers.

[1] Thinking philosophically and biologically, your realities—i.e., the manifestation of the union of numerous fundamental phenomena—can be *too complex to know or understand*. Specifically, your realities encompass your physiological realities, neuroscientific realities (pertaining to conscious and subconscious minds), psychological realities, and sociological realities.

Contents

Acronyms

ABC	Applied and basic combined
AI	Artificial intelligence
Co-PI	Co-principal investigator
DIKW	Data, information, knowledge, and wisdom
GPS	Global Positioning System
Ph.D.	Philosophy of doctorate
PI	Principal investigator
SED	Science, engineering, and design
STEM	Science, technology, engineering, and mathematics
w.r.t.	with respect to

Part I
Setting the Stage for a Ph.D.

Part I records two chapters on setting the stage of a Ph.D.:

- **Chapter 1: What Is a Ph.D.?**
- **Chapter 2: Why Are Pitfalls Inevitable in a Ph.D.?**

Chapter 1
What Is a Ph.D.?

A thousand mile journey begins with one step.

—Lao Tzu

Abstract A Ph.D. entails independent and original research, contribution to the candidate's field, a doctoral project of sufficient difficulty and scope, and mastery of writing and communication skills. Because these requirements present multi-faceted challenges, you shall first answer the following questions before embarking on a Ph.D. study: Why do you want to do a Ph.D.? Are you ready to stay committed for 3–5 years? Are you ready for any "failure"? How much are you prepared for a Ph.D.? To help you answer these questions, this chapter discusses the requirements of a Ph.D., the realities and myths of a Ph.D., the different forms of scholarship, the contemporary Ph.D. systems, the objectives of a Ph.D., generic checklist for Ph.D. success, readiness assessment for a Ph.D., and the career prospects of a Ph.D. graduate.

Keywords Ph.D. systems · Ph.D. requirements · The realities and myths of a Ph.D. · The objectives of a Ph.D. · Readiness assessment for a Ph.D. · Forms of scholarship

1.1 Introduction

Read not to contradict and confute, nor to believe and take for granted. . .but to weigh and consider.

—Francis Beacon (1561–1626)

The contemporary degree structures of colleges and universities have their roots in more than 700 years of tradition, and so is the association of advanced degrees with written theses and dissertations [1, p. xv]. As a degree of an academic discipline in 1861, the *Doctor of Philosophy (Ph.D.)* degree was first offered in the United States by Yale University, which was followed in 1890 by New York University, whose

T. M. Getu, *Pathways for Inspired Scientists and Engineers*,
https://doi.org/10.1007/978-3-032-06784-5_1

then Graduate School of Pedagogy offered the Ph.D. plus a *Doctor of Pedagogy* degree, which was credited as the first doctoral-level degree in education—a professional discipline—that was awarded in the United States [1, p. xvi-xvii]. Long before such a formal Ph.D. system, artisans and craftsmen systematized the preparation of new specialists by imposing a sequence of training—while protecting their livelihoods and skills from becoming everyone's property—from apprenticeship to the status of master, which marked the successful completion of training, after the preparation of one's masterpiece that was adjudicated to be worthy of the name by a jury of master's [1, p. xvi]. Inspired by this model, the thesis and dissertation became the pinnacles of successive levels of achievement, as academic centers emerged and the tradition of study evolved [1, p. xvi]. The master's and the doctorate then became distinguishing symbols [1, p. xvi].

During the twenty-first century, many universities across the globe have launched various Ph.D. programs featuring one or more of the contemporary Ph.D. systems—such as an essential Ph.D., a master's plus Ph.D., an industrial Ph.D., a professional Ph.D., and an interdisciplinary Ph.D.—according to societal and industry needs, funding availability, their research strength and focus, their research strategy, and the funding agencies' investment strategies. Despite these diverse factors determining a graduate school's choice of a Ph.D. system and its program focus, a Ph.D. entails successfully demonstrating the Ph.D. requirements, which are often intellectually challenging to any graduate student. The Ph.D. requirements that enforce a high standard to any Ph.D. student's intellectual journey are independent research, original research, novel contributions to the candidate's research field, a doctoral project of sufficient difficulty and scope, and mastery of the writing and communication skills needed to broadly disseminate one's research results [1, 2], which all have to be demonstrated by a Ph.D. student who is expected to research under and collaborate with a Ph.D. (co-)supervisor as *hand in glove* while being fully independent. Such an intellectual journey is fraught with multifaceted challenges due to the inherent uncertainty associated with conducting original research that is coupled with a Ph.D. student's lack of adequate research and publication experience, leading to the inevitability of Ph.D. pitfalls in managing oneself and one's Ph.D.

The unavoidable Ph.D. pitfalls are generic pitfalls that can happen during a Ph.D. or specific pitfalls that can temporarily occur during the initial, middle, or final years of a Ph.D. study. During the initial and middle years of a Ph.D., any Ph.D. student can likely come across numerous persona-related pitfalls, management-related pitfalls, research-related pitfalls, writing and publication-related pitfalls, and motivation-related pitfalls, which are temporary pitfalls that would indeed disappear as a Ph.D. student keeps working *creatively and persistently hard* toward the final year of a Ph.D. study, when a Ph.D. student can likely face various completion-related pitfalls and management and funding-related pitfalls. These are all temporary specific Ph.D. pitfalls, whose detrimental impact will disappear as one becomes mature in one's Ph.D. research. However, succeeding as a Ph.D. student also demands resilience, persistence, constant learning, and thoughtfulness in overcoming several generic Ph.D. pitfalls, such as maintaining a work-life balance, intensive engagement in life, expectations (your expectations and your supervisor's), maintaining your curiosity, a lack of comprehensive research skills, and developing as a critical thinker, a creative thinker, a strategic thinker, a probabilistic thinker, a causal thinker, and an independent thinker. These pitfalls can occur throughout one's Ph.D. journey, and their alleviation requires diverse success strategies, which shall be deployed by a persistent, curious, and hardworking mind.

Given the specific and generic Ph.D. pitfalls that are likely to happen in anyone's Ph.D. journey, an aspiring undergraduate or graduate student shall genuinely answer the following questions before pursuing a Ph.D. study: Why do you want to do a Ph.D.? What do you need a Ph.D. for? Do you need a Ph.D. badly (as a demonstration of who you are and what you want to do)? What do you expect from yourself as a Ph.D. student who should be committed for 3–5 years of your life? Are you going to be *fearless* for a wild scholarly journey? Are you ready to accept any "failure"? Are you ready to move forward in the face of multifaceted challenges? How much are you ready for a Ph.D.? To help you answer these questions and establish a bridge between the completion of a graduate study and the inception of a Ph.D. study, this chapter offers critical discussions on the several aspects of a Ph.D. study and research, such as the requirements of a Ph.D., the realities and myths of a Ph.D., the different forms of scholarship, the contemporary Ph.D. systems, the objectives of a Ph.D., the generic checklist for Ph.D. success, the readiness assessment for a Ph.D., and the career prospects of a Ph.D. graduate.

The remainder of this chapter is organized as follows. Section 1.2 details the requirements of a Ph.D. Section 1.3 provides the realities and myths of a Ph.D. Section 1.4 highlights the different forms of scholarship. Section 1.5 discusses the contemporary Ph.D. systems. Section 1.6 delineates the generic and specific objectives of a Ph.D. Section 1.7 outlines the generic checklist for a Ph.D. success. Section 1.8 discusses readiness assessment for a Ph.D. Section 1.9 highlights the career prospects of a Ph.D. graduate. Finally, Sect. 1.10 ends this chapter with concluding remarks.

1.2 The Requirements of a Ph.D.

> *Strength and growth come only through continuous effort and struggle.*
>
> *—Napoleon Hill (1883–1970)*

According to Marian Petre and Gordon Rugg, who are the authors of a book entitled *The Unwritten Rules of PhD Research* [2], a Ph.D. is a demonstration of one's competence in research as evidenced by a Ph.D. dissertation that establishes research insight, respect for one's discipline, mastery of one's subject, capacity for independent research, and ability to communicate one's results and narrate them to the broader discourse [2, p. 2].

A Ph.D. student is expected to demonstrate one's capacity for independent research in a chosen field by publishing top-tier journals, magazines, and conference papers. A Ph.D. student is expected to demonstrate research insight and mastery of one's topic via each of one's published journal papers addressing a specific research challenge or proposing a new solution, by researching a chosen field in depth. A Ph.D. student is expected to demonstrate the ability to communicate one's results and detail them to the broader discourse by relating each result in a publication and a Ph.D. dissertation. A Ph.D. student is expected to demonstrate respect for one's chosen discipline through novel research works developed by researching in depth, possibly in breadth too.[1]

By working hand in glove with a Ph.D. (co-)supervisor while consistently demonstrating independence, a Ph.D. student's research activities are anticipated to culminate in a Ph.D. dissertation. According to James E. Mauch and Namgi Park, who are the authors of a book entitled *Guide to the Successful Thesis and Dissertation: A Handbook for Students and Faculty* [1], a Ph.D. dissertation must qualify at least for a minimum intellectual standard, as remarked below [1, p. 3–4].

> The doctoral dissertation is expected to represent independent and original research in the field of the candidate's graduate study. It must add, in some fashion, to understanding in the candidate's field. Such contribution to knowledge may result either from the critical examination of materials not hitherto dealt with or from the re-examination of traditional materials by means of new techniques or from new points of view. The project undertaken must be of sufficient difficulty and scope to test the candidate's ability to carry on further research [independently] and it must ensure . . . mastering the skills needed for such research.

Accordingly, *five criteria for a standard Ph.D. dissertation* are outlined below.

! Attention: Five Criteria for a Standard Ph.D. Dissertation

- **Independent research**: Starting from the inception of one's Ph.D. research, a Ph.D. student is expected to demonstrate one's skills in conducting independent research, which is ought to be complemented by the insight and expertise of a Ph.D. (co-)supervisor so that one can enhance the overall quality of the work.
- **Original research**: A Ph.D. student is expected to make novel cutting-edge contributions that can qualify for the high standards of top-tier journals, which may assess originality as a new theory or a reinterpretation of an existing theory, a new or improved product, a new or improved model, a new perspective, a new or improved research tool (or technique), a critical analysis, an in-depth study, research-based portfolio of work, a fact or conclusion (or a collection of facts or conclusions), and an exploration of a topic, area, or field [3, p. 199].

[1] Since the research focus of a Ph.D. student depends on the type and number of research projects investigated by a Ph.D. (co-)supervisor as a principal investigator (PI) or Co-PI, delivering a variety of contributions and compiling them into a Ph.D. dissertation can be acceptable, as directed by some Ph.D. (co-)supervisors who encourage contributions in breadth.

- **Contribution to the candidate's research field**: As a part of one's intellectual journey in a chosen research field, a Ph.D. student is expected to deliver considerable knowledge contributions, which one's Ph.D. dissertation will document.
- **Sufficient difficulty and scope**: While independently conducting original research, a Ph.D. student is expected to engage with a Ph.D. project of sufficient difficulty and scope, so one's ability to carry out further independent research is tested.
- **Mastering the skills**: To engage with a Ph.D. project of sufficient difficulty and scope, a Ph.D. student is expected to master all the necessary skills needed to conduct original research independently.

If a Ph.D. student has to *begin with the end in mind* and work backward to what has to be done every day, one must persistently target the itemized five criteria to accomplish a standard Ph.D. dissertation.

Many Ph.D. (co-)supervisors often advise Ph.D. students to publish at least three journal papers that would be a foundation for a manuscript-based Ph.D. dissertation or a monograph. If a Ph.D. dissertation comprises several unpublished chapters, its expected quality should satisfy the minimum standard set by the UK Research Councils' joint statement quoted below [3, p. 257].

> PhD students are expected to make a substantial, original contribution to knowledge in their area, normally leading to published work. The development of wider employment-related skills should not detract from that core objective.

Specifically, a Ph.D. student is expected to demonstrate multiple abilities, self-thinking, comprehensive understanding, and more profound knowledge. Regarding multiple abilities, a Ph.D. student is expected to demonstrate one's ability to recognize and validate problems; develop theories; critically analyze and assess findings of oneself and others; and summarize, document, report, and reflect on one's research progress [3, p. 257]. On self-thinking, a Ph.D. student is expected to demonstrate original, independent, and critical thinking [3, p. 257]. Concerning a comprehensive understanding, a Ph.D. student is expected to demonstrate comprehensive understanding of relevant research methodologies and techniques along with their applications in one's research field [3, p. 257]. Regarding more profound knowledge, a Ph.D. student is expected to demonstrate deeper knowledge of recent advances in one's research field and in its related areas [3, p. 257]:

Because a Ph.D. is awarded as a recognition of one's status as an independent researcher, effective learning of high-level skills, and successful implementation of those skills [4], below is a perspective on why the process of obtaining a Ph.D. is as important as the products of a Ph.D. dissertation themselves [4, p. 8].

> Indeed, a strong implication of the emphasis on becoming a professional researcher is that the process of the PhD (learning and practice of high-level research skills) is at least as important as the product of the PhD (the research findings in the thesis).

As a Ph.D. student, you shall demonstrate you will be a thoroughly professional researcher, regardless of your career goals. According to Estelle M. Phillips and Derek S. Pugh, who are the authors of a book entitled *How to Get a PhD: A Handbook for Students and their Supervisors* [5], here is a perspective on what a Ph.D. is for [5, p. 59]:

> You are not doing some research for its own sake; you are doing it in order to demonstrate that you are a fully professional researcher, with a good grasp of what is happening in your field and capable of evaluating the impact of new contributions to it – your own as well as others'. That is what you get the doctorate for.

We now move on to our discussion on the realities and myths of a Ph.D.

1.3 The Realities and Myths of a Ph.D.

> *Sometimes life seems a dark tunnel with no light at the end, but if you just keep moving forward, you will end up in a better place.*
>
> *—Jeffrey Fry*

This section presents a consensus on the realities and myths of a Ph.D. while emphasizing important arguments advocated by existing books on how to get a Ph.D.

1.3.1 The Realities of a Ph.D.

> *No matter how gloomy the morning, it is a new day where anything is possible.*
>
> *—Jeffrey Fry*

The widely accepted realities of a Ph.D. are highlighted below for the awareness and inspiration of current and prospective Ph.D. students.

☛ **Emerging with a Ph.D. Will Make Anyone a Stronger and Better Person**
Through numerous opportunities for critical thinking, creative thinking, strategic thinking, probabilistic thinking, causal thinking, and independent thinking, a Ph.D. will make anyone a better person [6] and thinker, changing you for the better regardless of your discipline and career goals.

☛ **A Decent Ph.D. Is Made of Top-Tier Journal or Conference Papers**
In many scientific and engineering fields, a decent Ph.D. is made of top-tier journal articles; in computer science, it is made of top-tier conference papers [6].

☛ **A Ph.D. Is Not "to Change the World nor to Make an Earth-Shattering Contribution"**
According to Sonja K. Foss and William Waters, who are the authors of a book entitled *Destination Dissertation: A Traveler's Guide to a Done Dissertation* [7], the function of a Ph.D. dissertation is not "to change the world nor to make an earth-shattering contribution" to your chosen research field [7, p. 4].

☛ **A Ph.D. Can Differ from a University to a University**
Widely attributed to the fact that there are many ways of doing the same thing, a Ph.D. can differ from a university to a university, so is from a Ph.D. (co-)supervisor to a Ph.D. (co-)supervisor, for various institutions' unique culture and research engagement.

☛ **A Ph.D. Candidate's Achievements Can Shape One's Future Career**
A Ph.D. candidate's achievements can shape one's future career in science/engineering, underscoring the need to deliver important contributions during a Ph.D. study.

☛ **Flunking a Ph.D. by Overestimating or Underestimating Its Requirements**
If a Ph.D. student fails to understand the nature of a Ph.D. by overestimating or underestimating its requirements, one can likely fail in getting a Ph.D. [5, Ch. 4].

☛ **A Ph.D. Student Can Always Succeed by Working Hand in Glove with a Ph.D. (Co-)Supervisor**
If a Ph.D. student continues to lose contact with a Ph.D. (co-)supervisor for any reason, one's likelihood of not getting a Ph.D. increases [5, Ch. 4].

☛ **A Ph.D. Entails Hard Work and Persistence Until the Last Minute**
A Ph.D. study undoubtedly requires hard work and persistence until the last minute, irrespective of the field.[2]

1.3.2 *The Myths of a Ph.D.*

> *The only person you are destined to become is the person you decide to be.*
>
> *—Ralph Waldo Emerson (1803–1882)*

For the information and awareness of current and prospective Ph.D. students, we hereunder highlight some of the widely believed myths of a Ph.D.

[2] Taking a new job before defending a Ph.D. dissertation can increase the probability of failure in a doctoral study [5, Ch. 4].

☛ **Doing a Ph.D. Is Coming Up with a Groundbreaking Discovery**
According to Marian Petre and Gordon Rugg, who are the authors of a book entitled *The Unwritten Rules of PhD Research* [2], a Ph.D. is all about coming up with a significant contribution, despite a mediocre one. Although there is no limit to the novelty of an innovation delivered by a Ph.D. candidate, the argument "doing Ph.D. is coming up with a groundbreaking discovery" is arguably the myth of a Ph.D.

☛ **Only Those Who Are "Very Smart" Can Earn a Ph.D.**
In the very dynamic and rapidly advancing scientific frontiers of the twenty-first century, the one who is "very smart" in advancing a given research theme may very well be dull for a rapidly evolving and new research theme, should one fail at catching up the rapid advancements of science and technology. Accordingly, everything boils down to *burning desire* and *persistence*, regardless of any situation or system, such as a Ph.D. While being mindful of the many pitfalls of a Ph.D., we firmly believe that anyone who has successfully made it to a graduate school can earn a Ph.D. as long as one has the burning desire and persistence to make it happen, underscoring that "those who are 'very smart' can earn a Ph.D." is the myth of a Ph.D. and possibly an overestimation of one's worth.

☛ **Ph.D. Research Is an "Ivory Tower" Activity**
Among the widespread misconceptions about research, such as Ph.D. research, the belief that research is an "ivory tower" activity—removed from reality and social interaction with others—is a popular myth [5, p. 17] worth debunking! Because the twenty-first century's pressing societal, technological, and global challenges necessitate novel interdisciplinary, multidisciplinary, and transdisciplinary research, which undoubtedly requires close collaboration among many experts from diverse research fields in stark contrast with the view of "ivory tower" activity, "Ph.D. research is an 'ivory tower' activity" is certainly the myth of a Ph.D.

☛ **Good Personal Relationships as a Substitute for Hard Work**
Although it is always advisable to forge a good personal relationship—that may facilitate a good working relationship—with a Ph.D. (co-)supervisor, such a personal relationship cannot substitute the hard work needed to satisfy the Ph.D. requirements, as discussed in Sect. 1.2. Therefore, "good personal relationships as a substitute for hard work" is the myth of a Ph.D.

☛ **Teamwork as a Substitute for an Independent Work**
Depending on the culture of a given research laboratory and the nature of its research projects, a Ph.D. (co-)supervisor may direct a Ph.D. student to engage in collaborative teamwork. However, one shall be mindful that teamwork is not a substitute for independent work, underlining the entitled pitfall, and a Ph.D. is awarded only for original work conducted independently [5, p. 18]. Hence, "teamwork as a substitute for an independent work" is the myth of a Ph.D.

We now move on to our discussion on the various forms of scholarship.

1.4 Forms of Scholarship

Science is a way of thinking much more than it is a body of knowledge.

—Carl Sagan (1934–1996)

According to the *Carnegie Association for the Advancement of Teaching* [8], the four forms of scholarship are *the scholarship of discovery, the scholarship of integration, the scholarship of application, and the scholarship of teaching*. These forms naturally exhibit a circular relationship, with each benefiting from the others while also contributing to them [9, p. 20–21], as highlighted below.

☛ **The Scholarship of Discovery**

Usually thought of as traditional research [9], the scholarship of discovery is aimed at discovering the governing principles, fundamental theories, fundamental asymptotic and non-asymptotic limits, rigorous algorithms, and fundamental tradeoffs. This form of scholarship benefits from the scholarship of integration, the scholarship of application, and the scholarship of teaching while contributing to all three.

☛ **The Scholarship of Integration**

Being not just multidisciplinary but interdisciplinary, the scholarship of integration makes connections across disciplines [9], partly as a response to the demands of industries that predominantly deal with problems unbounded by specific disciplines [9]. This form of scholarship benefits from the scholarship of discovery, the scholarship of application, and the scholarship of teaching while also contributing to all three.

☛ **The Scholarship of Application**

The scholarship of application comprises service directly tied to a researcher's special field of expertise [9]. This form of scholarship benefits from the scholarship of discovery, the scholarship of integration, and the scholarship of teaching while also contributing to all three.

☛ **The Scholarship of Teaching**

Through the education and training of prospective scholars, the scholarship of teaching is aimed at imparting, transforming, and extending knowledge [9]. This form of scholarship benefits from the scholarship of discovery, the scholarship of integration, and the scholarship of application while also contributing to all three.

As for the scholarship of integration, a dual-scholarship paradigm dubbed *applied and basic combined (ABC) scholarship* is emerging by amalgamating the scholarship of discovery and the scholarship of application [10], as emphasized below.

☛ **ABC Scholarship**

According to Ben Shneiderman, who is the author of a book entitled *The New ABCs of Research: Achieving Breakthrough Collaborations* [10], combining basic and applied research yields higher-impact research as opposed to separately doing them [10, p. 1], provided that a researcher deploys the *science, engineering, and design (SED)* principle that blends the methods of science, engineering, and design [10, p. 1–2].

We now continue with our discussion on the contemporary Ph.D. systems.

1.5 The Contemporary Ph.D. Systems

There is nothing either good or bad but thinking makes it so.

—William Shakespeare (1564–1616)

According to the changing societal and industry needs, funding availability, and funding agencies' investment strategies,[3] the twenty-first-century global research institutions and universities host numerous Ph.D. programs that focus on different Ph.D. systems, such as *an essential Ph.D., a master's plus Ph.D., an industrial Ph.D., a professional Ph.D.*, and *an interdisciplinary Ph.D.* [2], which we henceforth discuss along with their *pros and cons*. We start our discussion with an essential Ph.D.

1.5.1 Essential Ph.D.

Enthusiasm moves the world.

—Arthur Balfour (1848–1930)

As detailed in Sect. 1.2, an essential Ph.D. is a conventional Ph.D. system that encourages minimum supervision for a Ph.D. student to have maximum independence and flexibility throughout a Ph.D. study [2], wherein a Ph.D. student is primarily expected to deliver a substantial innovation and novelty. This Ph.D. system is mainly pursued for the scholarship of discovery and the scholarship of integration within the scope of a Ph.D. research project, consistent with the aims of traditional research. In the initial phase of this Ph.D. system conducted in North America, a Ph.D. student must attend and pass advanced courses—upon the recommendation of a Ph.D. (co-)supervisor—in the intended Ph.D. research project's areas.

Ideal for anyone interested in pursuing an academic career, an essential Ph.D. is a widely accepted traditional Ph.D. system, which is crucial for an early-stage research career of independent discovery and innovation. However, it inherently has its own cons. First, a student of this Ph.D. system is mainly geared toward an academic career, making oneself inadequately ready for an industry career or entrepreneurship. Second, although a graduate of this Ph.D. system may emerge

[3] Funding agencies' investment strategies determine the carrier trajectories and focus of many principal investigators (PIs).

as a promising scholar in one's research field, one tends to have narrow training, which may not be rigorous for non-academic careers. Third, a graduate of this Ph.D. system may have limited collaboration and professional experience, as the system expects them to be completely independent and innovative.

1.5.2 Master's Plus Ph.D.

Sometimes when I consider what tremendous consequences come from little things, I am tempted to think there are no little things.

—Bruce Barton (1886–1967)

Common in North America, a master's plus Ph.D. system is conducted to ensure that a Ph.D. student demonstrates a comparable degree of competence and sophistication in one's research field—for a year or two—before one is sent off into the Ph.D. phase of autonomy and independent research, reassuring a Ph.D. student that one has the competence for conducting independent research [2]. This Ph.D. system has advanced Ph.D. courses that must be passed in a year or two, after which a master's degree is granted by some institutions to the successful students who would be promoted to a Ph.D. study [2]. Oftentimes, an elevated graduate student pursues the lines of research one would be conducting throughout a master's program, should this program be a master's by research, making a Ph.D. student undergo an essential Ph.D. system after one has advanced into a Ph.D. study. Accordingly, this Ph.D. system has the pros and cons of an essential Ph.D. system as highlighted in Sect. 1.5.1, apart from its mentioned pros.

1.5.3 Industrial Ph.D.

Dreamers only dream, but creators bring their dreams into reality.

—Robert Fritz (1943-)

This Ph.D. system focuses on applied research of industrial interest and entrepreneurial promise. A Ph.D. student pursuing industrial Ph.D. is usually expected to embed one's Ph.D. research in an organization's research while working for it and drawing on its resources [2]. While the organization directs the work, the university sets the academic standards of this Ph.D. system, which is often led by a university-based Ph.D. supervisor in collaboration with an industry-based supervisor [2]. A Ph.D. student pursuing this Ph.D. system is expected to undertake challenging industrial problems—unbounded by specific disciplines—and conduct novel scholarship of integration and scholarship of application.

Since industrial and academic perspectives can sometimes be in non-negotiable conflicts regarding priorities, deadlines, and intellectual property, a Ph.D. student pursuing an industrial Ph.D. can feel isolated despite the richness of the industry-academy environment and its significant opportunities [2]. A Ph.D. student pursuing an industrial Ph.D. may not likely have the significant experience and achievements of an independent scholarship of discovery, contrary to graduates of an essential Ph.D. Eventually, a graduate student pursuing an industrial Ph.D. may have difficulties to become academically productive, as demonstrated by the quality and quantity of top-tier publications, since focusing on two things at once—i.e., bridging academic and industrial perspectives by solving challenging industrial problems—can be difficult for many graduate students with little to no prior research experience.

1.5.4 Professional Ph.D.

Success is often the result of taking a misstep in the right direction.

—Al Bernstein (1950-)

To recognize a significant level of professional skill than a master's degree, professional doctorates have been established in many universities since the 1990s [5, p. 196].

Professional doctorates are intended for professionals who wish to combine research with professional practice, recognizing that professional domain expertise can contribute to research expertise, because professional practice itself offers a relevant skill set and context for research in the professional domain [2, p. 23]. As much as a Ph.D. requires a contribution to the advancement of one's chosen academic research field, a professional Ph.D. requires a candidate to deliver contributions that advance one's professional practice [5, p. 196].[4]

A professional Ph.D. program often involves preliminary research activities and a final project, which are conducted under the guidance of a supervisory team of an academic from a university and a practitioner from an organization for which the professional application takes place [2, 5]. In the preliminary research phase, a professional Ph.D. student typically engages with advanced taught courses, surveying the latest developments in one's professional research field and developing a professional research proposal, and alike, which must be satisfactorily completed before a professional Ph.D. candidate advances to the final project that often comprises the application of professional knowledge and skill to the solution of research problems in a real-world setting [5, p. 197]. The successful execution of this project would lead to the key component of the final thesis, which must

[4] If a Ph.D. is for a *professional scholar*, a professional Ph.D. is for a *scholarly professional* [5, p. 196].

demonstrate an independent self-analysis of the completed work and reflection on the use of academic knowledge in a practical setting [5, p. 197]. Accordingly, a successful graduate of a professional Ph.D. program will be a skilled professional who has not only gathered practical experience but also investigated the use of one's academic and practical knowledge, which sums up a contribution to a professional practice [5, p. 197].

1.5.5 Interdisciplinary Ph.D.

> *If you want to succeed you should strike out on new paths, rather than travel the worn paths of accepted success.*
>
> —*John D. Rockefeller (1839–1937)*

Apart from the scholarship of discovery that is often the goal of an essential Ph.D., it is possible to make original contributions through the scholarship of integration, the scholarship of application, or the scholarship of teaching [9, p. 353], which is according to Richard M. Reis who authored a book entitled *Tomorrow's Professor: Preparing for Academic Careers in Science and Engineering* [9], calling for a Ph.D. system designed to make doctoral students deliver original contributions to knowledge via the three forms of scholarship. Such a Ph.D. system is called an interdisciplinary Ph.D., which aims for the scholarship of integration or the scholarship of application while training Ph.D. students to solve practically useful and challenging interdisciplinary problems that cannot be addressed by a single discipline alone [9]. An interdisciplinary Ph.D. is becoming popular due to many funding agencies' current strategic focus on interdisciplinary research, which is widely believed to have a promising potential to address the numerous societal and technological challenges of this twenty-first century.

If one's doctoral training maintains the rigor of an essential Ph.D., a Ph.D. graduate's career prospects can be enhanced by an interdisciplinary Ph.D. This Ph.D. system is better suited to the needs of particular institutions, which are prepared to reward interdisciplinary scholarship in their faculty retention, promotion, and tenure process [9, p. 353–354]. On the contrary, according to Karen L. Kelsky, who is a professor-turned-academic career coach with the brand "The Professor Is In,"[5] an interdisciplinary Ph.D. can decrease a recipient's career prospects for there are fewer interdisciplinary departments than disciplinary ones, and institutional focus and strategy on interdisciplinarity are not matched by administrative and organizational support of interdisciplinarity [11]. Nevertheless, regarding the impactfully ongoing *science and technology convergence* [12], not only an interdisciplinary Ph.D. but also a *multidisciplinary Ph.D.* and a *transdisciplinary Ph.D.* are poised to be common in many research universities, consistent with the grand challenges

---

[5] Available at https://theprofessorisin.com/.

of twenty-first-century life and economy calling for convergence-inspired bold solutions.

Besides five Ph.D. systems discussed, there are Ph.D. study models, such as a *project-based Ph.D.*, a *Ph.D. by publication*, and a *part-time Ph.D.* [2, p. 21–24], with their own pros and cons. Therefore, each discussed Ph.D. systems can be *a project-based Ph.D.*, *a Ph.D. by publication*, or *a part-time Ph.D.*, according to the discretion of a Ph.D. (co-)supervisor, funding availability, and institutional rules.

We now proceed with our discussion on the objectives of a Ph.D.

1.6 The Objectives of a Ph.D.

The only source of knowledge is experience.

—Albert Einstein (1879–1955)

Around the globe, there are various Ph.D. systems and specifics that aim to achieve different Ph.D. objectives. First, different Ph.D. supervisors may have different convictions on what the objectives of a Ph.D. should be, as influenced by their respective Ph.D. research experience and achievements while inspired by their different research and professional experience garnered from different academic enterprises. Second, doctoral students may have a different perception regarding the objectives of a Ph.D. study, so those who wish to pursue an academic career may have a different perception from those who wish to pursue a non-academic career, underscoring the different commitment and research attitude of the latter and former group of students. Third, universities are inclined to make their Ph.D. programs geared to specific goals inspired by their research excellence, strategic plan, and research strength.

Toward unifying the objectives of all parties participating in a Ph.D. program, we hereunder discuss the generic and specific objectives of a Ph.D.

1.6.1 The Generic Objectives of a Ph.D.

We are what we repeatedly do. Excellence, then, is not an act, but a habit.

—Aristotle (384-322 BC)

According to James E. Mauch and Namgi Park, who are the authors of a book entitled *Guide to the Successful Thesis and Dissertation: A Handbook for Students and Faculty* [1], Ph.D. students, university faculties, and higher education institutions have their generic objectives—as the leading players of Ph.D. programs—regarding Ph.D. programs they all participate in [1], as remarked below.

☛ **The Generic Objectives of Ph.D. Students**
The generic objectives of Ph.D. students are professional and academic standing, completing course work at a high-quality level, staying within the statute of limitations, and finding good advisors and role models [1, p. 27–29].

☛ **The Generic Objectives of University Faculties**
Faculties aim to advance their scholarship by seeking the truth, building on the existing body of knowledge, and publishing cutting-edge works [1, p. 30].

☛ **The Generic Objectives of Higher Education Institutions**
Higher education institutions aim at providing leadership, expanding knowledge, and improving professional practice [1, p. 29].

1.6.2 The Specific Objectives of a Ph.D.

Light tomorrow with today!

—Elizabeth Barrett Browning (1806–1861)

To satisfy the generic objectives of a Ph.D., a Ph.D. program shall have the following *educational objectives* [13].

! Attention: The Specific Objectives of a Ph.D.

Lifelong self-learning capability, logical reasoning capability, capability of conducting independent and innovative research, writing skills, communication skills, oral presentation skills, networking skills, and miscellaneous skills.

Next, we briefly discuss each of the mentioned specific objectives.

☛ **Lifelong Self-Learning Capability**
Since Ph.D. is the highest degree one could get, there will be nobody to teach one after graduating with a Ph.D. A Ph.D. student shall thus develop a lifelong self-learning capability to survive in a constantly changing technological world, for there are always new technological trends, practices, and innovations [13].

☛ **Logical Reasoning Capability**
A Ph.D. student shall develop a logical reasoning capability by acquiring skills for *deductive reasoning* that enables one to plausibly reason from the general to the special; *inductive reasoning* that empowers one to credibly reason from the special to the general; and *abductive (plausible) reasoning* that qualifies one to rigorously identify the hidden factors that cause the observed phenomena [13].

☛ **Capability of Conducting Independent and Innovative Research**
A Ph.D. student shall develop and demonstrate a capability of conducting innovative and independent research, regardless of one's research topic [13].

☛ **Writing Skills**
A Ph.D. student shall acquire impressive writing skills to publish high-quality journal and conference papers, magazine papers, book chapters, and a Ph.D. dissertation, apart from grant proposals [13].

☛ **Communication Skills**
A Ph.D. student shall forge advanced communication skills—in both writing and speaking—that would make one's ideas and views crystal clear to others [13].

☛ **Oral Presentation Skills**
A Ph.D. student shall develop appreciable oral presentation skills that would engage, inspire, and enlighten audiences at conferences, seminars, and scientific events [13].

☛ **Networking Skills**
A Ph.D. student shall cultivate considerable networking skills to effectively socialize with other Ph.D. students, researchers, professionals, industry sponsors, government agencies, and not-for-profit organizations [13].

☛ **Miscellaneous Skills**
A Ph.D. student shall carve out essential leadership and managerial skills, such as visionary, inspiring leadership, and effective management [13].

We now proceed with our discussion on the generic checklist for Ph.D. success.

1.7 Generic Checklist for Ph.D. Success

Success is not built on success. It's built on failure. It's built on frustration. Sometimes it's built on catastrophe.

—Sumner Redstone (1923–2020)

In an open-ended and globally distributed competitive ecosystem such as Ph.D. research enterprise, achieving success can be tough, since everyone is also looking for it. However, because being a *principle-centered Ph.D. student* is a viable way to manage any uncertainties that arise down the invisible scientific roads, you—as a Ph.D. student—need to manage yourself and your Ph.D. effectively. Accordingly, we hereinafter outline *success checklists for managing yourself and your Ph.D.*

1.7.1 Success Checklist for Managing Yourself

That some achieve great success is proof to all that others can achieve it as well.

—Abraham Lincoln (1809–1865)

The underneath is a generic checklist for managing yourself as a Ph.D. student.

! Attention: Success Checklist for Managing Yourself

Staying motivated, effective time management, managing distractions, stress management, being on standby, full engagement with your Ph.D. research, and living Peter's Laws.

In what follows, each list is highlighted.

☛ **Staying Motivated**

To drive success in your Ph.D. study, you must always stay motivated, regardless of any pitfalls you may encounter during your study.

☛ **Effective Time Management**

You shall routinely apply a *2D time management* (considering *urgency and importance*) or a *3D time management* (considering *urgency, importance, and significance*).

☛ **Managing Distractions**

Because you are expected—as a Ph.D. student—to spend most of your time researching and thinking, you shall always manage distractions that can easily occur, such as unnecessary phone calls, invitations, social life, and research breaks, apart from moments of insufficient concentration.

☛ **Stress Management**

Since the uncharted scholarly roads of a Ph.D. compounded by your lack of research experience—as a mere graduate student—can cause stress anytime, you must manage stress reactively, whenever it arises, and proactively.

☛ **Being on Standby**

As a Ph.D. student, you shall often be on standby, since there can be constant paper submissions, paper reviews, ideation, experimentation, and simulation.

☛ **Full Engagement with Your Ph.D. Research**

Since you can advance your knowledge and imagination by being more absorbed with your Ph.D. research, you shall always be fully engaged with your Ph.D. research.

☛ **Living Peter's Laws**

As a Ph.D. student, you may live *Peter's Laws* [14] because they undoubtedly inspire the cognitive development of your persistent and passionate mind.

1.7.2 Success Checklist for Managing Your Ph.D.

> *The person interested in success has to learn to view failure as a healthy, inevitable part of the process getting to the top.*
>
> *—Joyce Brothers (1927–2013)*

The following is a generic checklist for managing your Ph.D.

! Attention: Success Checklist for Managing Your Ph.D.

Becoming unconventionally curious, working creatively and hard, being too proactive, learning all the time, boundless imagination, asking the right questions, and firm research ethics.

Next, we highlight every list for your guidance and inspiration.

☛ **Becoming Unconventionally Curious**

As a curious Ph.D. student, you are not supposed to believe all the existing ideas, algorithms, and claims, which are always opportunities for unbounded curiosity. Thus, you may ask why all the time; wonder about the mathematical proofs, claims, and cited references; contemplate a possible extension of any published paper; and spot the weaknesses and limitations in any published work.[6]

☛ **Working Creatively and Hard**

Because a Ph.D. student's success is inextricably linked to one's number of executed experiments or simulations, whose outputs are often uncertain and unknown in advance, nobody exactly knows the optimal effort needed to earn a Ph.D. As a Ph.D. student, therefore, you should always *work creatively and hard*.

☛ **Being Too Proactive**

Because success in often uncharted territory—such as Ph.D. research—is correlated with proactivity, a Ph.D. student must surely be proactive all the time.

☛ **Learning All the Time**

Since there are countless things to learn in depth and breadth, a Ph.D. student shall surely learn all the time while creating and seizing learning opportunities.

☛ **Boundless Imagination**

Since imagination is more important than knowledge, as affirmed by Albert Einstein, a Ph.D. student needs to imagine relentlessly; fantasize about great science, ideas, methods, and solutions; aspire to think out of the box all the time; and persistently envisage to stand on the shoulders of the giants.

☛ **Asking the Right Questions**

Since questions drive research throughout [15], a Ph.D. student must keep asking *the right questions*, which can lead to novel research directions.

☛ **Firm Research Ethics**

As a Ph.D. student of firm professionalism and character, you shall always give credit where it is due; wholeheartedly recognize others' endeavors as much as you want yours to be recognized; never plagiarize or engage in any act of plagiarism, including self-plagiarism; and never engage in any act or collaboration that may make you succeed in the short run but hurt your scientific journey in the long run.

We now move on to briefly discuss readiness assessment for a Ph.D.

---

[6] As a curious researcher, you are not supposed to be a conformist. Hence, you may question everything; disagree respectfully, even with your Ph.D. (co-)supervisor and renowned researchers; disprove any existing theory; and be inspired by an existing theory to develop a new one. As you become increasingly curious, you will realize that there are astronomical things waiting to be discovered, unlike your possible initial feeling that most discoveries have already been made.

1.8 Readiness Assessment for a Ph.D.

Perseverance is not a long race; it is many short races one after the other.

—Walter Elliot (1888–1958)

How much will you be ready to pursue a Ph.D. study that would take 3–5 years? To offer useful perspectives, we start by providing the following questions that are essential to evaluate your readiness for a Ph.D. study.

! Attention: Questions for Assessing Your Readiness for a Ph.D.

- *Are you ready to commit 3–5 years of your life to your Ph.D.?*
- *Are you ready to stay motivated regardless of the inevitable Ph.D. pitfalls?*
- *Will you condition your mind to remain motivated for 3–5 years?*
- *Are you willing to persistently and happily learn?*
- *Are you ready to have research among your genuine hobbies?*
- *Are you ready for "failures?"*

Underneath, we provide perspectives concerning each itemized question.

☛ Are You Ready to Commit 3–5 Years of Your Life to Your Ph.D.?

Because a Ph.D. study needs a lot of hard work and determination, this essential hard work may force many Ph.D. students to hold on to their personal lives and hobbies for multiple years.

☛ Are You Ready to Stay Motivated Regardless of the Inevitable Ph.D. Pitfalls?

Many Ph.D. students lose the initial motivation that draws them to a Ph.D. study, which requires them to be persistent for an extended time, as long as 3–5 years. However, a Ph.D. study requires staying motivated throughout the years, even during some unpleasant times that are inevitable due to many Ph.D. pitfalls.

☛ Will You Condition Your Mind to Remain Motivated for 3–5 Years?

Succeeding in a Ph.D. entails years of persistent hard work, executed by a motivated mind that may need to be conditioned for 3–5 years.

☛ Are You Willing to Persistently and Happily Learn?

To conduct novel research as a Ph.D. student, you need to learn numerous cutting-edge concepts, theories, and practices of your research field almost daily.

☛ Are You Ready to Have Research Among Your Genuine Hobbies?

A possibly best strategy for enduring Ph.D. research with minimum pain is to have research among your genuine hobbies, regardless of your success in research.

☛ Are You Ready for "Failures?"
Since you—as a Ph.D. student and new researcher—have limited research experience, knowledge, and expertise in a given field, it is common to "fail"[7] a number of times during a Ph.D. study, especially when you attempt to conduct novel research.

Summarizing, according to Carlo Ghezzi, who is the author of a book entitled *Being a Researcher: An Informatics Perspective* [16], the following is a list of critical abilities that are potential indicators for being fit for research [16, p. 76–77].

! Attention: Potential Indicators for Being Fit for Research [16, p. 76–77]

Intellectual curiosity, openness to new ideas, willingness to learn more, attraction to engaging with non-trivial problems, persistence in problem solving, ability to critically assess and go beyond the apparently obvious answers, rigor, ambition, resilience to failure, ability to abstract and formalize, analytical skills, mathematical skills, and ability to collaborate.

We now proceed with the career prospects of a Ph.D. graduate.

1.9 The Career Prospects of a Ph.D. Graduate

The future influences the present just as much as the past.

—Friedrich Nietzsche (1844–1900)

Although a Ph.D. is generally believed to make its recipient overqualified, a Ph.D. degree can open many doors in academia, government, business, and non-profit, affording rich opportunities for *conventional, unconventional, and hybrid careers* toward your bright future. A Ph.D. degree offers many opportunities for conventional careers at academic institutions, industries, and government laboratories, such as national labs. A Ph.D. degree also offers many opportunities for unconventional careers such as commercialization of your research, entrepreneurship, authorship, freelancer-ship, inception of a not-for-profit organization, earning MBA and pursuing management, and consultancy, which are possible due to the numerous skills you can acquire by doing a Ph.D. A Ph.D. degree also offers many opportunities for hybrid careers—that are plausibly attractive for early-stage research scholars—such as post-doc and entrepreneurship, post-doc and government research, and government research and entrepreneurship. Summing up, since your Ph.D. journey

[7] Many motivational authors and thinkers underscore that there are *no "failures," but wrong results.*

shall be a part of your larger intellectual journey toward a post-Ph.D. bright future, a Ph.D. offers numerous post-Ph.D. career opportunities to be created and seized, for which you shall be ready in advance while creating time.

1.10 Concluding Remarks

The key to growth is the introduction of higher dimensions of consciousness into our awareness.

—Lao Tzu

As part of setting the stage for this book and Ph.D. research, this chapter sought to establish a bridge between the completion of a graduate study and the inception of a Ph.D. study. For this purpose, we provided important discussions on the several aspects of a Ph.D. study, such as the requirements of a Ph.D., the realities and myths of a Ph.D., the different forms of scholarship, the contemporary Ph.D. systems, the generic and specific objectives of a Ph.D., generic checklist for a Ph.D. success, readiness assessment for a Ph.D., and the career prospects of a Ph.D. graduate.

Although we tried to present crucial perspectives on what a Ph.D. is and what it takes to earn one, our limited discussion—due to a space constraint—may not have covered the *full-blown depth and breadth* of issues intertwined with the discussed themes. For additional insights and realities, we encourage all Ph.D. students to reflect this chapter's references and discuss them with their network of peers.

References

1. Mauch JE, Park N (2003) Guide to the Successful Thesis and Dissertation: A Handbook for Students and Faculty, 4th edn. Dekker, New York, NY
2. Petre M, Rugg G (2010) The Unwritten Rules of PhD Research, 2nd edn. Open University Press, Berkshire
3. Cryer P (2006) The Research Student's Guide to Success, 3rd edn. Open University Press, Berkshire
4. Finn JA (2005) Getting a PhD: An Action Plan to Help Manage Your Research, Your Supervisor and Your Project. Routledge, New York, NY
5. Phillips EM, Pugh DS (2005) How to Get a PhD: A Handbook for Students and their Supervisors, 4th edn. Open University Press, Berkshire
6. Guo PJ (2015) The Ph.D. Grind: A Ph.D. Student Memoir. Available http://www.pgbovine.net/PhD-memoir.htm. Cited May 2025
7. Foss SK, Waters W (2016) Destination Dissertation: A Traveler's Guide to a Done Dissertation, 2nd edn. Rowman & Littlefield, Lanham
8. The Carnegie Foundation (2020) The Carnegie Foundation for the Advancement of Teaching. Available https://www.carnegiefoundation.org. Cited May 2025
9. Reis RM (1997) Tomorrow's Professor: Preparing for Academic Careers in Science and Engineering. IEEE Press, New York, NY

10. Shneiderman B (2016) The New ABCs of Research: Achieving Breakthrough Collaborations. Oxford University Press, Oxford (Kindle Edition)
11. Kelsky KL (2020) The Professor Is In: The Curse of the Interdisciplinary Ph.D. Available https://chroniclevitae.com/news/548-the-professor-is-in-the-curse-ofthe-interdisciplinary-ph-d. Cited May 2025
12. Bainbridge WS, Roco MC (eds.) (2016) Handbook of Science and Technology Convergence. Springer, Switzerland
13. Wu DO (2020) Advise on creative thinking, research, writing and speaking. Available http://www.wu.ece.ufl.edu/knowhow.html. Cited May 2025
14. Diamandis PH (2024) Peter's Laws. https://www.diamandis.com/peters-laws
15. Snieder R, Larner K (2009) The Art of Being a Scientist: A Guide for Graduate Students and their Mentors. Cambridge University Press, New York, NY
16. Ghezzi C (2020) Being a Researcher: An Informatics Perspective. Springer, Cham

Chapter 2
Why Are Pitfalls Inevitable in a Ph.D.?

Growth and comfort do not coexist.

—Ginni Rometty (1957-)

Abstract As a Ph.D. student, you are expected to create a new *knowledge network* through your prospective Ph.D. dissertation that would be linked to its references. Since the state-of-the-art knowledge network varies over time and every Ph.D. dissertation addresses a unique research problem(s), the corresponding knowledge network is unique, making the dissemination of any Ph.D. dissertation fundamentally indeterministic. This translates to various Ph.D. pitfalls throughout your Ph.D. study. Accordingly, this chapter underscores the inevitability of Ph.D. pitfalls, the determinants of your Ph.D. experience, the tradeoffs of a Ph.D., why you will surely succeed with your Ph.D., and a specific checklist for your smooth Ph.D. journey.

Keywords Ph.D. pitfalls · The determinants of your Ph.D. experience · Tradeoffs of a Ph.D. · Checklist for a smooth Ph.D. journey

2.1 Introduction

Change is the end result of all true learning.

—Leo Buscaglia (1924–1998)

Your Ph.D. dissertation is expected to demonstrate mastery of your subject and research insight through innovative scholarly works that are independently conducted and effectively communicated to the broader research communities [1, p. 2], who naturally expect authors to demonstrate mastery and respect for the research discipline. As a Ph.D. student, you are expected to create new knowledge that is linked to the broader state-of-the-art knowledge and advancements through your Ph.D. dissertation, which constitutes a new *knowledge graph* (*knowledge network*) where the dissertation is a node connected to several other nodes that are the published works cited by it.

T. M. Getu, *Pathways for Inspired Scientists and Engineers*,
https://doi.org/10.1007/978-3-032-06784-5_2

As to your Ph.D. dissertation's origination of a *citation network*, the links among the dissertation and the cited publications are the interconnecting concepts that are the foundations of the dissertation. As everyone's Ph.D. dissertation addresses *unique research problem(s)* in a possibly peculiar way, the corresponding knowledge network is also unique, making the process of disseminating a knowledge network unique and *indeterministic* for every Ph.D. student. This indeterminism is attributed to the fact that the broader state-of-the-art knowledge network varies substantially over time, rendering a time-varying number of nodes and interconnecting links that make every Ph.D. student's experience in authoring a Ph.D. dissertation unique. Therefore, the overall process of pursuing a Ph.D. is indeterminate and has some inherent randomness.

To succeed toward your Ph.D. and beyond, you must develop various skills and capabilities that would enable you to conduct independent and innovative research. The significant development of such skills and capabilities as early as possible can make your Ph.D. journey smooth, rewarding, and entertaining. Meanwhile, according to Phil Agre, who is the author of a book entitled *Networking on the Network: A Guide to Professional Skills for PhD Students* [2], you will face an irreducible *chicken-and-egg problem* [2, Section 7, p. 41][1] during the early stages of your Ph.D. study. To this end, you need to be part of a research community to do research; on the contrary, you cannot be part of a research community without doing research [2, Section 7, p. 41]. This chicken-and-egg problem is generally at its worst in the middle years of your Ph.D., so the entire globe can seem chaotic, and your candidate research themes will seem impossibly huge [2, Section 7, p. 41]. What you will similarly encounter at different times are the unforeseen Ph.D. difficulties and hence *Ph.D. pitfalls*.

In light of the chicken-and-egg problem and the inherent randomness of a Ph.D., the structural challenges are the recurring Ph.D. pitfalls that can hinder you from progressing in your Ph.D. research, unless you continually acquire and apply various skills to overcome them. Interestingly, these pitfalls would disappear[2] as if they were never there beforehand, provided that you keep working hard to consistently grow in your research field while acquiring and applying various Ph.D. skills. Your skills acquisition can be effectively informed by advanced awareness of the Ph.D. pitfalls that will happen at different phases of your Ph.D. study, so you can be ready to surmount them effectively whenever they arise. It is crucial to know the overall Ph.D. pitfalls that can happen during the different phases of your Ph.D. study. Accordingly, detailed discussions on the many Ph.D. pitfalls—inevitable during the initial, middle, and final years of a Ph.D. study—are worth providing.

---

[1] This problem is also known as *the chicken-or-egg problem* and *the chicken-versus-egg problem*.

[2] With a high probability close to 1.0, the inevitable Ph.D. pitfalls will vanish, provided that you keep developing the essential Ph.D. skills and habits, necessitating you to be a curious reader who must not draw negative energy out of the Ph.D. pitfalls, which are opportunities.

Since accomplishing novel Ph.D. research amounts to *motivated and thoughtful intellectual journey amid pitfalls,* anything can happen during your Ph.D. study, exactly like sailing in an unsettled weather, echoing the indeterminism inherent in pursuing a Ph.D. study. The indeterminism can be amplified during the initial years of your Ph.D. study, which is impacted by the various uncertainty factors such as the uncertainty associated with the transition from a master's study to a Ph.D. study. Like everyone's unique life experience that can also be affected by many uncertainty factors, each Ph.D. student goes through a unique scholarly journey, leading to a unique Ph.D. experience. Nevertheless, almost all Ph.D. students inevitably face many similar Ph.D. pitfalls during the initial, middle, and final years of their Ph.D. studies. Therefore, your entire Ph.D. journey shall be thoughtfully and purposefully executed with a *motivated mind* that is well aware of the Ph.D. pitfalls in advance.

The remainder of this chapter is organized as follows. Section 2.2 highlights why the beginning of a Ph.D. study is uncertain. Section 2.3 presents the factors that determine your Ph.D. experience. Section 2.4 highlights the tradeoffs of a Ph.D. Section 2.5 follows with the reasons for the inevitability of Ph.D. pitfalls. Section 2.6 argues why you will undoubtedly succeed in an intellectual endeavor toward a Ph.D. degree. Section 2.7 outlines a specific checklist for your smooth Ph.D. journey. Finally, Sect. 2.8 ends this chapter with concluding remarks.

2.2 Why Is the Beginning of a Ph.D. Uncertain?

> *The only thing that can be predicted is the probability of different events.*
>
> *—Richard P. Feynman (1918–1988)*

At the beginning of your Ph.D. study, you would transition from being a graduate student to a research student who is expected to manage both coursework and research work effectively. This multitasking likely complicates the initial years of your Ph.D. study to the extent that your beginning becomes highly uncertain, requiring your complete understanding of the underlying beginning-phase uncertainty factors of your Ph.D. study. These factors are—but not limited to—itemized below.

! Attention: The Beginning-Phase Uncertainty Factors of Your Ph.D. Study

- *You are expected to manage both coursework and research work.*
- *You may work in a new research field.*
- *You likely have no prior work to build upon.*
- *You may be new to the supervision style of your Ph.D. (co-)supervisor.*
- *You have no or limited experience working with a Ph.D. (co-)supervisor.*
- *You likely have inadequate publication experience.*
- *You may not know the required effort to get your paper accepted.*

- *You may not know the level of novelty enforced by a top-tier journal or your Ph.D. (co-)supervisor.*
- *You must convince the academic gatekeepers to get it published.*

The itemized uncertainty factors are highlighted below.

☛ **You are Expected to Manage Both Coursework and Research Work**
During the initial years of your Ph.D. study, you are expected to manage coursework and research work concurrently, in line with the expectation of Ph.D. (co-)supervisors in many North American and some European/Asian universities.[3] Although managing two things at once is undoubtedly an opportunity when both succeed, it is cognitively challenging, as our brains can only focus on one thing at a time. Managing both coursework and research work is hence an uncertainty factor.

☛ **You May Work in a New Research Field**
Since your Ph.D. project is directed according to the available funding of your Ph.D. (co-)supervisor, you may end up working on a new research field whose state-of-the-art advancements and research landscape can be entirely new to you, underscoring an uncertainty factor.

☛ **You Likely Have No Prior Work to Build Upon**
Unless you are advancing your graduate research in the framework of a *master's plus Ph.D. program*, you likely have no prior advanced work to build on. This reality underscores the necessary commitment to developing substantial subject-matter expertise, which is required to be a published author, entailing an uncertainty factor.

☛ **You May Be New to the Supervision Style of Your Ph.D. (Co-)Supervisor**
You will not know one's supervision style and expectations unless you have worked with your Ph.D. (co-)supervisor before. Since managing and adapting chemistry, besides character, can be challenging, this often constitutes an uncertainty factor.

☛ **You Have No/Limited Experience Working with a Ph.D. (Co-)Supervisor**
As a new Ph.D. student, you will be new to your Ph.D. (co-)supervisor's working style, writing style, personal preference, perspectives, and liking/disliking, which can entail a magnificent uncertainty factor (also a risk factor).

☛ **You Likely Have Inadequate Publication Experience**
Unless you are exceptional and fortunate enough to publish journal papers during your master's study, you have no prior or adequate publication experience, which is undoubtedly both a risk factor and an uncertainty factor.

☛ **You May Not Know the Required Effort to Get Your Paper Accepted**
Unless you publish some journal papers before starting your Ph.D. study, you may not know the persistent effort required to get your paper accepted. New to a research career, the typically high-bar endeavor may put you in a *zero/rejected position*

---

[3] In many North American and some European/Asian universities, a Ph.D. study entails mandatory coursework during its initial years. In contrast, in some other European universities, there is no mandatory coursework, though a Ph.D. student may attend some according to their research needs.

regardless of your substantial effort, which is undoubtedly an uncertainty and risk factor.

☛ **You May Not Know the Level of Novelty Enforced by a Top-Tier Journal or Your Ph.D. (Co-)Supervisor**
In a top-notch research enterprise, persistent hard work is a requirement for a beginning Ph.D. student, but never a sufficient one. Hence, you may not be able to accurately estimate the level of novelty expected by your Ph.D. (co-)supervisor or a top-tier journal, affirming a novelty-related uncertainty factor.

☛ **You Must Convince the Academic Gatekeepers to Get It Published**
After working on and concluding a paper, you must convince the academic gatekeepers to publish it in a decent journal or conference. As an outsider to the community of anonymous reviewers who are active researchers and subject-matter experts, you would be new to its writing styles, preferences, scholarly judgment, and technical scrutiny, asserting an uncertainty as well as a risk factor.

All the aforementioned uncertainty factors, along with how you manage yourself, your research, and your Ph.D. (co-)supervisor, will largely determine your Ph.D. experience, as discussed below.

2.3 The Determinants of Your Ph.D. Experience

Reality is an acquired taste.

—Robert Fritz (1943-)

Depending on your academic and research background, relationship with Ph.D. (co-)supervisor, and readiness to undertake scholarly challenges, you will come across many specific and generic Ph.D. pitfalls, whose management will directly or indirectly determine your overall Ph.D. experience. To this end, the main determinants of your Ph.D. experience are—but not limited to—the following.

! Attention: The Main Determinants of Your Ph.D. Experience

- *Your research background and experience.*
- *Your enrolling Ph.D. university.*
- *Prior research experience related to your Ph.D. project.*
- *Prior publication experience related to your Ph.D. project.*
- *The type of your Ph.D. (co-)supervisor.*
- *Your professional relationship with your Ph.D. (co-)supervisor.*
- *The research group you are working with.*
- *Different work culture and research style.*
- *The type of research field you work in.*
- *The types of conference and journal papers you will be publishing.*

- *The number and type of active researchers in your research field.*
- *Ambition, motivation, and goals.*

The itemized determinants are highlighted below.

☛ **Your Research Background and Experience**

Your research experience and background—gained in your bachelor's and master's studies—determine your Ph.D. experience and engagement, especially during the initial years of your Ph.D. study.

☛ **Your Enrolling Ph.D. University**

Your Ph.D. university determines your Ph.D. experience through its doctoral requirements and regulations, Ph.D. committee members, pre-doc requirements, research focus and strength, research tradition, minimum graduation expectancy, and so forth.

☛ **Prior Research Experience Related to Your Ph.D. Project**

Prior research experience related to your Ph.D. project can determine your Ph.D. experience during the initial years, especially in managing a dual workload stemming from coursework and research work. However, it is neither necessary nor sufficient.

☛ **Prior Publication Experience Related to Your Ph.D. Project**

Previous publication experience related to your Ph.D. project can determine your Ph.D. experience, especially your publishing experience during the initial years of your Ph.D., though it is neither necessary nor sufficient.

☛ **The Type of Your Ph.D. (Co-)Supervisor**

Your Ph.D. (co-)supervisor directly determines your Ph.D. experience via their expertise, funding, behavior, research engagement, expectation, scholarly challenges and scrutiny,[4] writing style, research style and focus, collaboration, research preference, work style, etc.

☛ **Your Professional Relationship with Your Ph.D. (Co-)Supervisor**

Because we humans are *neuroscientifically subjective creatures*, your Ph.D. experience can also be shaped by your professional relationship with your Ph.D. (co-)supervisor, which is easy to break but challenging to develop. Accordingly, delivering excellent results that demonstrate your research excellence, scientific courage, and scientific curiosity is the primary factor that helps you develop a long-lasting and rewarding relationship with your Ph.D. (co-)supervisor.[5]

---

[4] Rising to the scholarly challenges and scrutiny of a Ph.D. (co-)supervisor is an intellectually rewarding activity you should enjoy, since challenges are always opportunities!

[5] If you encounter issues in your relationship with your Ph.D. (co-)supervisor, you should be thoughtful enough to see things from both your and your (co-)supervisor's perspectives; see the "*Employer's Viewpoint*" in [3, p. 95–98].

☛ **The Research Group You Are Working With**
The research group you are working with directly determines your Ph.D. experience via its research collaboration, guidance and support, relationship with you and your Ph.D. (co-)supervisor, scientific ambition, scientific courage, research engagement, research environment, and alike.

☛ **Different Work Culture and Research Style**
It is often heard and witnessed that Ph.D. students have their unique working culture and research style. Some tend to be *lone wolves*; some like being collaborators; some persist as frequent publishers; some are inclined to industry; some have academic inclination; and so forth [4]. Thus, your adopted working culture and research style determine your Ph.D. experience.

☛ **The Type of Research Field You Work In**
Any two Ph.D. students gather different Ph.D. experiences mainly due to the type of their research fields that afford unique research problems. Since the type of research field you work on directly shapes your Ph.D. experience, here is a question worth asking: *Is your Ph.D. research field a new, old, hot, or reviving one?* If you are engaged in a new research field, you will have the opportunity to come across many uncharted scholarly territories for novel intellectual contributions. However, there would be stiff competition among the researchers "fighting for" new resources and opportunities. If you are working on an old or a reviving research field, you would enjoy several support systems and resources from the preceding published researchers that have been doing a great deal of work in defining, analyzing, surveying, scrutinizing, and clearing the research roads. However, you may find it hard to deliver novel contributions because the research field's resources may be depleted.

☛ **The Types of Conference and Journal Papers You Will Be Publishing**
The types of journal papers you will publish hugely determine your Ph.D. experience through the journals' impact factor, acceptance rate, editors, anonymous reviewers, etc. The same is true for your conference papers, since they are usually the basis for journal papers.

☛ **The Number and Type of Active Researchers in Your Research Field**
The number and type of active researchers working in your research field can determine your Ph.D. experience, since they can be your reviewers; they can be the conference organizers and journal editors; they can be the ones who expect you to maintain the research standard of a community publishing in some prestigious journals; they can be the ones who compete with you for citations; and some of them may also set the research standard as visionary research leaders and thinkers.

☛ **Ambition, Motivation, and Goals**
Because your ambition, goals, and motivation can determine your Ph.D. experience and achievements, they are important determinants of doctoral experience that can pave the way to your successful career in science and innovation.

Depending on your comprehensive understanding of the discussed determinants, you may be able to estimate your prospective Ph.D. experience while also considering the tradeoffs of a Ph.D. that we discuss below.

2.4 The Tradeoffs of a Ph.D.

The price of excellence is discipline. The cost of mediocrity is disappointment.

—William Arthur Ward (1921–1994)

As the adage goes, life is a series of tradeoffs, and so is a Ph.D. whose tradeoffs you should be aware in advance. The following are—but not limited to—the tradeoffs of a Ph.D.

! Attention: The tradeoffs of a Ph.D.

Research life versus personal life, independence versus supervisor guidance, high-quality versus average-quality Ph.D. research, publish more versus publish less, high-quality versus average-quality Ph.D. dissertation, and certain versus uncertain research future.

These tradeoffs are argued below.

☛ **Research Life Versus Personal Life**

Compared with your graduate research experience, the high expectations and research standards of a Ph.D. translate into many hours of daily research work, which must be executed with a motivated and persistent mind. This intensive research engagement can transform your personal life into a research life, which you may live as your hobby, forcing you to give up on your personal life. When you focus on your work-life balance and allocate more time for your personal life, your Ph.D. research can significantly suffer. Therefore, there is always a tradeoff between your research life and personal life, which is one of the realities of being a Ph.D. student. Armed with this knowledge, you shall be determined enough not to be bogged down by intensive engagement in your personal life, which is a generic Ph.D. pitfall.

☛ **Independence Versus Supervisor Guidance**

As a Ph.D. student, your Ph.D. (co-)supervisor usually expects your intellectual independence, which you must demonstrate throughout your Ph.D. study while rising to its challenges daily. Irrespective of any challenge, you shall be proactive, adaptive, and resilient enough to independently deal with your research problems, since your Ph.D. (co-)supervisor does not necessarily have the time to deal with the intricacies of your Ph.D. research. While dealing with your research complexities more independently, you depend less on your Ph.D. (co-)supervisor. However, your research wisdom and experience are inadequate not to need your Ph.D. (co-)supervisor's guidance. Despite this fact, waiting for your Ph.D. (co-)supervisor's guidance makes you more dependent. Hence, there is an inherent tradeoff between your independence and the need for your Ph.D. (co-)supervisor's guidance, requiring a balance between independence and guidance.

☛ **High-Quality Versus Average-Quality Ph.D. Research**
High-quality theoretical Ph.D. research entails envisioning a novel idea; significant and persistent endeavors to realize the envisioned idea via mathematical tools and computer simulations/experiments; and demonstrating the significance of an algorithm born out of the envisaged idea. Realizing all these leads you to novel Ph.D. research published in a top-notch journal of your research field. On the other hand, if your envisioned idea is not novel enough, your necessary realizing efforts are not significant and persistent enough, or you fail to demonstrate the significance of your technique born out of your envisioned ideas, you will likely end up publishing average-quality Ph.D. research. Average-quality versus a high-quality Ph.D. research as a tradeoff boils down to whether you generate a novel idea that you would put substantial effort into realizing.

☛ **Publish More Versus Publish Less**
You can produce several first-author conference/journal papers throughout your Ph.D. study for persistent and adaptive hard work, substantial research commitment, and mastery of the intricacies of your research field. Once you creatively and persistently pursue numerous research directions, possibly at the same time, your many investigations fueled by your persistent endeavors can lead to your several first-author papers, which can make you competitive, provided that your research field values the quantity of papers. However, producing numerous incremental contributions may not make you produce high-impact breakthrough papers that can instantly garner numerous citations and wider readership,[6] which are crucial academic metrics that can immediately spell the quality of your work and your reputation as an independent researcher. Although you shall also target high-quality and even breakthrough papers that report on your possible discovery, a discovery often requires considerable effort and too much time, which is needed to develop significant expertise in your research field. Consequently, focusing on high-impact and breakthrough papers will result in fewer publications. Therefore, there is an inherent tradeoff between publishing more and publishing less, which is properly captured through a tradeoff named *quantity versus quality of papers*.

☛ **High-Quality Versus Average-Quality Ph.D. Dissertation**
Whether it is a manuscript-based Ph.D. dissertation or not, a high-quality Ph.D. dissertation is a collection of high-quality journal papers, often the consequence of original research and contributions. To author a high-quality Ph.D. dissertation, you must thus be delivering more novel Ph.D. contributions while relying less on the state-of-the-art literature and advancements of your research field. On the other hand, should you be satisfied with an average-quality Ph.D. dissertation, you may rely on your research field's state-of-the-art literature that may inspire you to make some incremental contributions. However, you are usually expected—

---

[6] Although authoring papers that garner numerous citations and wider readership shall be your ultimate goal in science, some Ph.D. advisors often advise—from realistic points of view—the Ph.D. students to focus on paper quantity at the early-stage Ph.D. research and paper quality at the late-stage Ph.D. research, which can be built on the already investigated research directions.

possibly by your Ph.D. (co-)supervisor and anonymous reviewers—to master each disseminated technique. Therefore, pertaining to incremental contributions and novel contributions, there is an inherent tradeoff between an average-quality Ph.D. dissertation and a high-quality Ph.D. dissertation.

☛ **Certain Versus Uncertain Research Future**

Everyone pursuing a Ph.D. seems to ponder a future research career in terms of a settled academic, government, or industrial research career. Because such settled research careers are usually in high demand, the realistic world has made it crystal clear that *being fiercely competitive* is the price that must be paid to achieve them. Certain research future thus entails extraordinary creativity, which you shall demonstrate throughout your Ph.D. study. On the other hand, if your Ph.D. study reveals that you were moderately curious and creative, an uncertain research future would likely and unfortunately be your cost. Hence, there is an inherent tradeoff between certain versus uncertain research future, which is captured via another tradeoff named *highly competitive versus averagely competitive Ph.D. creativity.*

It is a fact that all Ph.D. students confront different Ph.D. pitfalls during their Ph.D. necessitating our discussion on why pitfalls are inevitable in a Ph.D. study.

2.5 Why Are Pitfalls Inevitable in a Ph.D.?

Success is a lousy teacher. It seduces smart people into thinking they can't lose.

—Bill Gates (1955-)

The main reasons several pitfalls are inevitable in a Ph.D. study are—but not limited to—mentioned below.

! Attention: Why Pitfalls Are Inevitable in a Ph.D. Study

- *You may not possess the survival skills.*
- *You are always expected to be independent.*
- *You need to master many publications before contributing meaningfully.*
- *Spotting a scholarship opportunity can be tough for a beginning scholar.*
- *There is always a demand for hard work in the life of a Ph.D. student.*
- *A research job naturally forces you to accept and create challenges.*
- *Developing expertise in a dynamic field entails a persistent endeavor.*
- *Your anonymous reviewers are likely more experienced than you.*
- *The research community implicitly imposes some standards of novelty.*
- *The high expectations of top-tier journals translate to high intellectual standards.*
- *Researchers who set the highest research standard are extraordinary.*

The itemized reasons are briefly explained below.

☛ **You May Not Possess the Survival Skills**

Because you would be new to the entire research enterprise by the time you start your research career as a Ph.D. student, you may not have the survival skills or complete awareness of the "trading rules" [1].

☛ **You Are Always Expected to be Independent**

As part of the Ph.D. requirements, you are always expected to demonstrate your independence, because your Ph.D. (co-)supervisor is too busy to guide you through every *Ph.D. research turn and roundabout.*

☛ **You Need to Master Many Publications Before Contributing Meaningfully**

As a Ph.D. student, you are expected to thoroughly understand the depth and breadth of your research field's state-of-the-art advancements before starting any meaningful research.

☛ **Spotting a Scholarship Opportunity Can Be Tough for a Beginning Scholar**

Because the world has numerous senior researchers who aspire to publish on almost every research theme to grow in their research careers, spotting a scholarship opportunity can be tough for a beginning Ph.D. student, who hardly possesses the *scholarly lenses of an expert.*

☛ **There Is Always a Demand for Hard Work in the Life of a Ph.D. Student**

Although you may spend many months obtaining the evidence that you believe supports the claim you wish to make as a Ph.D. researcher, your evidence could unfortunately be insignificant in the mind of an expert researcher with many scholarly lenses, demanding extra persistent hard work that may take months.

☛ **A Research Job Naturally Forces You to Accept and Create Challenges**

Because conducting novel research is all about learning new things and thinking in novel ways to deliver new theories, algorithms, findings, and analyses, this very nature of a research job typically forces you to *accept and create challenges.*

☛ **Developing Expertise in a Dynamic Field Entails a Persistent Endeavor**

To develop considerable expertise at the level of your research field's leaders, you need to master not only your research field's state-of-the-art knowledge and advancements through an unrelenting endeavor over a long time but also constantly evolve your already developed expertise.

☛ **Your Anonymous Reviewers Are Likely More Experienced than You**

Since your prospective anonymous technical reviewers are likely to be more experienced than you, they may not be easily impressed by any incremental contributions and may become frustratingly demanding, which is a never-easy Ph.D. experience that parallels *moving uphill.*

☛ **The Research Community Implicitly Imposes Some Standards of Novelty**

When you attempt to solve a problem in your research field, the research community may implicitly impose some standards of novelty as your prospective anonymous technical reviewers, since this could be how it has gone through the "system."

☛ **The High Expectations of Top-Tier Journals Translate to High Intellectual Standards**
Since impact-oriented researchers overcome the limitations of the existing theories and algorithms by publishing advanced innovations in top-tier journals, the high expectations of top-notch journals implicitly enforce high intellectual standards.

☛ **Researchers Who Set the Highest Research Standard Are Extraordinary**
Because the highest standard of cutting-edge research is set by extraordinary researchers who are *persistent*, *proactive*, *visionary*, *dynamic*, *curious*, *independent and collaborative*, *passionate*, *creative*, *adaptive*, *highly effective*, *fearless*, and *bold*, enhancing their innovations or devising new science implicitly enforces essential scholarly qualities similar to the mentioned ones.

As argued below, despite the inevitability of many Ph.D. pitfalls, you have numerous reasons to succeed as a curious and hardworking Ph.D. student.

2.6 Why Will You Succeed Surely with Your Ph.D.?

> *Courage and perseverance have a magical talisman, before which difficulties disappear and obstacles vanish into air.*
>
> *—John Quincy Adams (1767–1848)*

The underneath are the reasons—but not limited to—why you will succeed surely with your Ph.D.

! Attention: Why You Will Succeed Surely with Your Ph.D.

- *Academia is a conversation about ideas, so you can contribute ideas to any conversation as well as anyone else.*
- *You can spark a magnificent motivation that can be sustained and amplified throughout your Ph.D.*
- *You will have a Ph.D. (co-)supervisor of thoughtful guidance as long as you demonstrate a keen interest and ambition.*
- *You can build your network over the network, which can be a support system.*
- *Millions of references will guide you in case of any confusion or questions.*
- *Things will get easier once you lay the Ph.D. foundation via your first contribution.*
- *You can be uniquely successful by thoroughly researching in depth and breadth.*
- *You can always discover many never-addressed research problems.*
- *You can always train your miraculous brain to be more intelligent.*
- *You can build infinite minds out of your 86 billion miraculous neurons.*
- *You are also an image maker, like every successful researcher.*
- *There are countless rooms for contributions so long as you are fearless and tirelessly imaginative.*

- *You have enormous possibilities for making new contributions.*
- *You can uniquely contribute, since everyone has a role in academia.*

All the itemized reasons are discussed below.

☛ **Academia Is a Conversation About Ideas, so You Can Contribute Ideas to Any Conversation as well as Anyone Else**
According to Sonja K. Foss and William Waters, who are the authors of a book entitled *Destination Dissertation: A Traveler's Guide to a Done Dissertation* [5], academia is a conversation about ideas [5, p. 6], so you can contribute ideas—as a curious new researcher—to any conversation as capably as anyone else.

☛ **You Can Spark a Magnificent Motivation That Can Be Sustained and Amplified Throughout Your Ph.D.**
Through your *ambition and expectancy*, you can spark a *high level of motivation* [6],[7] which can be sustained throughout your Ph.D. study by keeping your *attention* and *effort* on your Ph.D. research. You can also *amplify* your sustained motivation via a great research *attitude* and *environment*.

☛ **You Will Have a Ph.D. (Co-)Supervisor of Thoughtful Guidance as long as You Demonstrate a Keen Interest and Ambition**
So long as you demonstrate a keen interest and ambition in your work, you can always engage your Ph.D. (co-)supervisor, who would provide thoughtful guidance via ample scholarly experience and leadership while frequently producing something promising.

☛ **You Can Build Your Network Over the Network, Which Can Be an Invaluable Support System**
You can build your *network over the network* [2, 7], an invaluable support system and resource in case of challenges, which you shall always seize as opportunities.

☛ **Millions of References Will Guide You in Case of Any Confusion or Questions**
Whenever you have any confusion or question, there are millions of disciplinary, interdisciplinary, multidisciplinary, and transdisciplinary references—in the form of books, book chapters, journal papers, conference papers, workshop papers, and technical reports—that can offer you not only new ideas, perspectives, and knowledge and but also the *opportunity for imagination*. Hence, your Ph.D. job is all about efficiently discovering most of these references to get the ultimate guidance needed for your independent and novel research.[8]

---

[7] The book in [6] is a self-help book whose author has researched it for decades. Although the academic community may not necessarily respect self-help books for their insufficient evidence, we refer to [6] because we consider it as a useful resource featuring valuable advice and declarations.

[8] Once you succeed in efficiently discovering the existing references, you can connect each one *through imagination and hard work* to contribute beyond the state-of-the-art advancements.

☛ **Things Will Get Easier Once You Lay the Ph.D. Foundation via Your First Contribution**
Because your first contribution marks "a hole" in your research field's state-of-the-art advancements, which you can modify and expand to contribute more, it can serve you as a *foundation and compass direction* for your entire Ph.D. research. If you succeed in your first paper, you will likely obtain your second, third, fourth, and so forth papers, because you will have the momentum to contribute. Therefore, things will get easier once you lay the foundation for your Ph.D.

☛**You Can Be Uniquely Successful by Thoroughly Researching in Depth and Breadth**
As you keep thoroughly researching in the *depth and breadth* of your research field, you will have a high probability of generating your unique and novel ideas, since you would cognitively dwell on a much bigger *innovation space (intellectual space).*

☛ **You Can Always Discover Many Never-Addressed Research Problems**
As you keep working effectively and curiously on a specific theme of your Ph.D. research, you will likely uncover many never-addressed research problems, which offer you the opportunity for novel investigations through your prospective papers.

☛ **You Can Always Train Your Miraculous Brain to Be More Intelligent**
According to Robert Kowalski, who is the author of a book entitled *Computational Logic and Human Thinking: How to be Artificially Intelligent* [8], people can directly use *computational logic* to improve their intelligence [8, p. ix]. You can be more intelligent by training your brain broadly using the advances of artificial intelligence.

☛ **You Can Build Infinite Minds Out of Your 86 Billion Miraculous Neurons**
Because *the mind is a brain in action* [9, 10],[9] you have a mind for writing, a mind for driving, a mind for speaking a learned language, a mind for swimming, etc. Once you learn a new skill, a particular group of neurons in your brain *network and fire together* to serve you during its execution. Using your brain's *86 billion neurons*[10] and over *100 trillion synapses*, you can literally have infinite combinations of a group of neurons that would network and fire together during the execution of a given task. Therefore, you can train and develop literally countless minds, which certainly empower you to master astronomically many concepts and skills.[11]

☛ **You Are Also an Image Maker, Like Every Successful Researcher**
From a fundamental neuroscience standpoint, we are all *image makers* as to our brain's interpretation of audio-visual stimuli, knowledge representation, memory

[9] Although there are ongoing debates about the brain and the mind, we subscribe—for the purpose of this book—to the view that *the mind is a brain in action.*

[10] The human brain has, on average, 86 billion neurons and 85 billion nonneuronal cells [11, p. 10665].

[11] Despite some fundamental limits of our brains, such as *catastrophic forgetting*, *catastrophic interference*, and *synaptic plasticity* [12, 13], it is theoretically possible to develop countless minds that enable you to master astronomically many concepts and skills.

formation, and memory retrieval, which are evident during our interaction with the external world via our five senses. Since successful researchers are also image makers, you too can be one of them by making your images that you would work tirelessly to corroborate through scientific evidence—i.e., laboratory experiment, mathematical proof, computer simulation, or computer experiment.

☛ **There Are Countless Rooms for Contributions so long as You Are Fearless and Tirelessly Imaginative**
Although the astronomically many twenty-first-century publications may threaten to make a starting researcher feel that almost all important inventions have already been done, *too little has been achieved* given a far-sighted science and technology, especially regarding interdisciplinary, multidisciplinary, and transdisciplinary innovation. Accordingly, there are endless rooms for discovery as long as you are *fearless and tirelessly imaginative*, since there is no lack of frontiers [14, p. 9386].

☛ **You Have Enormous Possibilities for Making New Contributions**
You have countless possibilities for making contributions, which is very interesting, and here is a scenario on how. Considering that 200 overall parameters affect your Ph.D. research theme, and you wish to publish a paper that investigates the effect of 50 parameters at once, you would roughly have $C(200, 50)$[12]—i.e., combination of 50 out of 200—possibilities to devise a new contribution. Although there could be some (fundamental) challenges regarding the feasibility of a parameter combination possibility, it must be underscored that you have *countless possibilities* for delivering new contributions.[13]

☛ **You Can Uniquely Contribute, Since Everyone Has a Role in Academia**
As a specialist, a generalist, a pragmatist, a visionary, or a theorist, everyone has a *unique role to play* in a given research field, so you can uniquely contribute to your doctoral research field as long as you are curious and hardworking enough.

After assuring you that you can indeed succeed with your Ph.D., we now proceed to outline a specific checklist for a smooth Ph.D. journey.

2.7 Specific Checklist for a Smooth Ph.D. Journey

Believe you can and you're halfway there.

—Theodore Roosevelt (1858–1919)

Many researchers agree that doing a Ph.D. is an investment for 3–5 years, a common duration a Ph.D. student is expected to commit to their Ph.D. project. As for such an investment, there is some probability it will not only succeed but also fail, underscoring its inherent risk due to a lack of complete control over the uncertainty

---

[12] $C(200, 50)$ is computed as $\binom{200}{50} = \frac{200!}{50!(200-50)!}$, an exceedingly huge number!

[13] Unlike such combinatorial logic-based reasoning on the innumerable possibilities for making new contributions, the quality of a contribution can hardly be captured using the same logic.

factors of a Ph.D. study. Because being too strategic can minimize any risk, you may invest in a *strategic and rigorous preparation* before joining a Ph.D. program while conducting some pre-Ph.D. research that will inspire your Ph.D. journey ahead. Particularly, you may develop *invaluable scholarly skills and habits* to make your Ph.D. research life smooth and rewarding. As you mature intellectually by doing so, you will likely have a smooth Ph.D. journey toward your Ph.D. dissertation.

As a *specific checklist for a smooth Ph.D. journey*, the following scholarly skills and habits—but not limited to—can make your entire Ph.D. journey smooth.

! Attention: Checklist for a Smooth Ph.D. Journey

Sparking motivation, hardworking skills, budding research skills, writing skills, communication skills, time management skills, non-procrastination skills, networking skills, a habit of experimentation, believing in yourself, readiness for failure, going out of your comfort zone, optimism and positive thinking, resilience, patience, fighting spirit, personal mottoes, and doing research for the fun of it.

In what follows, we discuss each element of this checklist.

☛ **Sparking Motivation**

Because a Ph.D. demonstrates one's competence in research, your prospective Ph.D. dissertation must demonstrate independent research, original research, contribution to your research field, sufficient difficulty and scope, and mastery of the necessary research skills [1, 15]. To succeed at these, you shall spark your motivation for your Ph.D. study via *ambition and expectancy* [6].

☛ **Hardworking Skills**

Since every Ph.D. (co-)supervisor expects a Ph.D. student to meet multiple deadlines and journal revision due dates, a Ph.D. student shall work effectively hard, even by embracing boredom and laborious work while having a minimum interruption.[14] Because every workshop/conference paper, journal article, Ph.D. proposal, or Ph.D. dissertation is produced through persistent hard work, there is no substitute for it. Therefore, like speaking a learned language being a skill, you shall develop hardworking skills throughout your Ph.D. study.

☛ **Budding Research Skills**

As a Ph.D. student, you are expected to generate new ideas in your research field; formulate a new research problem in your research field; choose an existing mathematical framework or develop a new one; solve your formulated research

---

[14] For the sake of a sustained scholarly focus needed to produce new ideas and perspectives, you are advised to have a minimum interruption throughout your Ph.D. study, since your mind shall not waste much time returning to your Ph.D. life realities.

problem; validate your obtained solution using simulations or (computer) experiments; compare your validated solution with the existing ones, if any; write a journal paper (also a conference one); submit and revise your written journal paper; and publish your accepted journal article. This process will repeat itself in a new way whenever you aim to solve a new research problem and publish your obtained solution. To be adept at this process and succeed in your Ph.D., you shall acquire budding research skills before joining your Ph.D. program.[15]

☛ **Writing Skills**

Because writing and publishing journal and conference papers are the integral parts of a Ph.D. study, as highlighted above, you may write—possibly daily—throughout your Ph.D. study. Hence, you shall develop advanced writing skills throughout your Ph.D. study, because the quality of your writing is also correlated with your success rate in publications.

☛ **Communication Skills**

Throughout your Ph.D. study, you may need to regularly interact with graduate student colleagues, post-docs, your Ph.D. (co-)supervisor, your enrolling institution's administrative professionals, and researchers attending conferences. You may also need to network and interact with researchers at other institutions to collaborate and exchange research experience. The success of all these interactions undoubtedly contributes—directly or indirectly—to the progress of your Ph.D. research and management of yourself as a researcher. Therefore, it could not be overstated that communication is an *indispensable key*, calling for developing your communication skills throughout your Ph.D. study.

☛ **Time Management Skills**

Since there is often too much to do throughout your Ph.D. study to the extent that time would be very scarce, you may need to rely on an efficient time management scheme throughout your Ph.D. study. Accordingly, you shall develop a basic time management skill—that accounts for the *urgency, importance, and significance* of an activity/engagement—throughout your Ph.D. study.

☛ **Non-procrastination Skills**

As highlighted above, arriving at a journal paper—starting from generating your new idea—is a long process, because each step often takes considerable time and effort. Unless you are well motivated and ambitious, such a long way often evokes excuses in your mind, so you frequently end up engaging in procrastination, which can be aggravated by a non-essential engagement with friends and colleagues. Thus, you shall resist the urge of procrastination—regardless of your situation—to progress smoothly according to your research plan. Since resisting such an urge may be difficult, you must develop skills for avoiding procrastination throughout your Ph.D. study.

[15] Prior research experience in one's Ph.D. research areas can make one's Ph.D. journey smooth, especially during the initial Ph.D. years. However, it is neither necessary nor sufficient.

☛ **Networking Skills**
Because networking can bring you abundant information, crucial collaboration, support, encouragement, invaluable insight, and important research experience, you are always encouraged to have excellent networking skills. Hence, you shall develop networking skills throughout your Ph.D. study.

☛ **A Habit of Experimentation**
As highlighted above, you must naturally perform several simulations or (computer) experiments on a new idea to arrive at a publication. Whenever you have any new idea, you shall first experiment with its validity or use simulation software. As you engage with experimentation as a habit, your likelihood of arriving at a publication substantially increases. As a gratifying habit that can pave the way to quality and quantity of publications, a habit of experimentation can also make you optimistic, even when your experiments fail. Therefore, you shall develop a habit of experimentation throughout your Ph.D. study.

☛ **Believing in Yourself**
When you come across several novel journal papers, you may be overwhelmed by the creativity of the published senior authors, especially during your initial Ph.D. years. However, those novel papers are usually the direct consequence of intensive developments—building expertise, idea origination, problem formulation, proposed solutions, and experimentation—over the years; they may also be the previously rejected ones whose authors worked persistently hard to publish them. Since these experienced authors put years of effort into publishing those papers, why should not they overwhelm you, since you are just starting your scholarly journey? It is often the case that your brain can accomplish as much as the intensity of your belief in yourself, which is intertwined with your *neurological belief system*. Therefore, before and throughout your Ph.D. study, you shall believe in yourself to demonstrate your independent research.

☛ **Readiness for Failure**
Interestingly, some researchers strongly believe that you are not researching provided that you are constantly having success [16, p. 69]. Although we all researchers—as neuroscientifically subjective creatures—would like to see our research succeed all the time, a regular part of a curious researcher's life is "failure" (i.e., a wrong result), which must be embraced since it usually offers an opportunity to draw pragmatic lessons that can lead to progress. When you are absolutely ready for failure, you are likely to be *fearless*, which is the most differentiating trait of ideal researchers; you are also likely to be optimistic and calm even when things are not working. Because a Ph.D. usually is full of unexpected alleys and roundabouts, which can fail anyone, you should be ready for failure throughout your Ph.D. study.

☛ **Going Out of Your Comfort Zone**
As emphatically believed by almost all researchers, if novel research of significance is to be achieved, one must go out of one's *comfort zone*. As a Ph.D. student, you should leave your comfort zone, which could be a background concept or framework you are comfortable with. As you keep going out of your comfort zone since the initial years of your Ph.D. study, your likelihood of building deeper and broader expertise increases, since you will later be comfortable with those

concepts/frameworks that would make you uncomfortable during your initial Ph.D. years. Hence, while being ready for failure, you shall develop the habit of leaving your comfort zone throughout your Ph.D. study.

☛ **Optimism and Positive Thinking**

During your Ph.D. study, anything can happen, such as paper rejection, mathematical intractability, disagreement with a Ph.D. (co-)supervisor, frustration, exhaustion, etc. Overcoming these energy drainers certainly requires psychological makeup, a positive energy source. In this vein, *optimism and positive thinking*[16] can make you keep going by holding on to your initial Ph.D. plan. Accordingly, you shall develop the habit of optimism and positive thinking throughout your Ph.D. study.

☛ **Resilience**

As discussed before, several things can go unexpectedly wrong during one's Ph.D. study, since a Ph.D. student typically lacks adequate research experience. Once you go through the many uncertainties of a Ph.D. and successfully develop into an independent researcher worth recognizing with a Ph.D., you must be resilient throughout your Ph.D. study. Speaking of resilience, optimism can be a substantial contributing factor. While being optimistic, you shall thus develop the habit of resilience throughout your Ph.D. study.

☛ **Patience**

Because you may likely have a *subjective attachment* to the research problem you are investigating [1], from the vantage point of a Ph.D. student's psychology, you would likely wish to see your research succeed in every possible way. However, since conducting research is usually an undertaking of several investigations that may lead to a rigorous answer to a well-formulated research question, you may likely encounter things you do not expect beforehand, including completely wrong results. In this regard, part of the challenge would be to stay patient and go through it until your Ph.D. defense and the submission of your final Ph.D. dissertation. Therefore, you shall develop a habit of patience throughout your Ph.D. study.

☛ **Fighting Spirit**

In academia, the transition from generating a new idea to producing and validating new results to writing and publishing a journal paper has hardly been straightforward for theoretical and experimental researchers. Thus, you must fight to be where you want to be, whether you are a theoretical, or experimental Ph.D. researcher. In particular, a fighting spirit can be helpful for your persistent endeavor, which could lead to a badly needed solution. As everyone is out in the wild academic ecosystem silently fighting for their interests and accomplishments, interestingly, why not you?[17] Hence, you may develop a fighting spirit throughout your Ph.D. study.

---

[16] You also have to be *pessimistic* in order to be more frustrated so that you can work hard as a *laser-focused researcher*. Accordingly, your *experiential wisdom*—which you need to acquire over the years—will help you answer the following questions: *When shall I be optimistic? When shall I be pessimistic?*

[17] While we do not necessarily advocate fighting for everything, we recognize that your Ph.D. journey can be rewarding for a fighting spirit, provided that it is properly exploited.

☛ **Personal Mottoes**

Personal mottoes are often empowering during challenging times in one's Ph.D. study, which are likely to happen due to the many inevitable Ph.D. pitfalls. When things are not working in a Ph.D. world of experimentation, your mottoes would inspire you to work persistently hard. Speaking of personal mottoes, consider the following quote:

> **Important**
>
> *There will be plenty of time to rest in the grave.*
>
> *—Paul Erdös (1913–1996)*

If you make this quote one of your mottoes, your mind will remind you that there will be plenty of resting time in the grave whenever you feel tired or distracted. Through this motto, you can gain positive energy and motivation regardless of exhaustion and distraction. For the same reason, you may have personal mottoes for your Ph.D. research. Therefore, you shall have your mottoes throughout your Ph.D. study.[18]

☛ **Doing Research for the Fun of It**

Among the scholarly skills and habits we advocated above, the habit of doing research for the fun of it is the most important one. You shall conduct research not only to earn your Ph.D. but also to have fun. In this respect, you may ponder the following quote of a legendary physicist and winner of the 1965 *Nobel Prize in Physics.*

> **Important**
>
> *You see, I get so much fun out of thinking that I don't want to destroy this pleasant machine that makes life such a big kick.*
>
> *—Richard P. Feynman (1918–1988)*

When you do research for the fun of it, you would implicitly be ready for failure and become fearless, which are perfect recipes for outstanding research. Doing research for the fun of it also develops the hobby of independent thinking and experimentation,[19] which in turn leads to higher quality and quantity of publications. Therefore, you are immensely encouraged to do research for the fun of it throughout your Ph.D. study.

---

[18] For your fitting personal mottoes, *1001 motivational quotes* of great minds are featured in [17].

[19] If you decipher the beauty, ingenuity, and creativity behind "complicated equations and theorems," it will be too much fun to invent your own during your Ph.D. study.

2.8 Concluding Remarks

Only one who devotes himself to a cause with his whole strength and soul can be a true master. For this reason mastery demands all of a person.

—Albert Einstein (1879–1955)

The uncertain beginning typical of a Ph.D. study and your intellectual journey in a possibly uncharted territory make the various Ph.D. pitfalls inevitable. To provide several perspectives around this reality, we underlined why things are uncertain during the beginning of your Ph.D. study; presented the determinants of your overall Ph.D. experience; exposed the tradeoffs of a Ph.D.; presented the reasons why various Ph.D. pitfalls are inevitable throughout your Ph.D. study; detailed assuring reasons why you will undoubtedly succeed in an intellectual endeavor toward a Ph.D.; and outlined specific checklist for your smooth Ph.D. journey. With these discussions and perspectives, we aimed to offer you the big picture on the inevitability of Ph.D. pitfalls and why you will surely succeed with your Ph.D., irrespective of its pitfalls. The sooner you have a clearer big picture, the better off you will be without spending time and energy on trial-and-error adjustments.

Although we attempted to discuss the most important issues on the aforementioned themes, our limited discussion—due to a space constraint—may not have covered the *full-blown depth and breadth* of issues intertwined with the discussed themes. For detailed insights and realities, we encourage you to reflect this chapter's references and discuss them with your network of peers.

References

1. Petre M, Rugg G (2010) The Unwritten Rules of PhD Research, 2nd edn. Open University Press, Berkshire
2. Agre P (2002) Networking on the Network: A Guide to Professional Skills for PhD Students, 2nd edn. University of California, Los Angeles, CA
3. Feibelman PJ (2011) A PhD Is Not Enough!: A Guide to Survival in Science, revised edn. Basic Books, New York, NY
4. Rosei F, Johnston T (2006) Survival Skills for Scientists. Imperial College Press, London
5. Foss SK, Waters W (2016) Destination Dissertation: A Traveler's Guide to a Done Dissertation, 2nd edn. Rowman & Littlefield, Lanham, Maryland
6. Burchard B (2014) The Motivation Manifesto: 9 Declarations to Claim Your Personal Power. Hay House, Carlsbad, CA
7. Levine AG (2015) Networking for Nerds: Find, Access and Land Hidden Game-Changing Career Opportunities Everywhere. Wiley, Hoboken, NJ
8. Kowalski R (2011) Computational Logic and Human Thinking: How to be Artificially Intelligent. Cambridge University Press, New York, NY
9. Dispenza J (2017) Becoming Supernatural: How Common People Are Doing the Uncommon. Hay House, Carlsbad, CA
10. Dispenza J (2012) Breaking the Habit of Being Yourself: How to Lose Your Mind and Create a New One. Hay House, Carlsbad, CA

11. Herculano-Houzel S (2012) The remarkable, yet not extraordinary, human brain as a scaled-up primate brain and its associated cost. Proc. Natl. Acad. Sci. 109:10661–10668
12. Parisi GI, Kemker R, Part JL, Kanan C, Wermter S (2019) Continual lifelong learning with neural networks: a review. Neural Netw 113:54–71
13. Bear MF, Conners BW, Paradiso MA (2016) Neuroscience: Exploring the Brain, 4th edn. Wolters Kluwer, Philadelphia, PA
14. Geman D, Geman S (2016) Science in the age of selfies. Proc Natl Acad Sci 113:9384–9387
15. Mauch JE, Park N (2003) Guide to the Successful Thesis and Dissertation: A Handbook for Students and Faculty, 4th edn. Dekker, New York, NY
16. Snieder R, Larner K (2009) The Art of Being a Scientist: A Guide for Graduate Students and their Mentors. Cambridge University Press, New York, NY
17. Vilord TJ (2004) 1001 Motivational Quotes for Success: Great Quotes from Great Minds. Garden State Publishing, Cherry Hill, NJ

Part II
Pitfalls of a Ph.D.

Part II documents five chapters on the pitfalls of a Ph.D.:

- **Chapter 3: Research-Related Pitfalls of a Ph.D.**
- **Chapter 4: Writing and Publication-Related Pitfalls of a Ph.D.**
- **Chapter 5: Management-Related Pitfalls of a Ph.D.**
- **Chapter 6: Persona and Motivation-Related Pitfalls of a Ph.D.**
- **Chapter 7: Generic Pitfalls of a Ph.D.**

Chapter 3
Research-Related Pitfalls of a Ph.D.

The best way out is to go through.

—Robert Frost (1874–1963)

Abstract Emanating from the indeterministic and challenging nature of a Ph.D. study, numerous specific and generic Ph.D. pitfalls are inevitable during the initial, middle, and final years of a Ph.D. study. The specific Ph.D. pitfalls are seasonable ones as diverse as research-related, writing and publication-related, management-related, and persona and motivation-related Ph.D. pitfalls. Among these pitfalls, this chapter discusses the various research-related Ph.D. pitfalls that are inevitable during the initial and middle years of your Ph.D. study while aiming to prepare you in advance so that you will not be bogged down by any research-related Ph.D. pitfalls.

Keywords Specific Ph.D. pitfalls · Generic Ph.D. pitfalls · Research-related Ph.D. pitfalls

3.1 Introduction

Good things do not come easy. The road is lined with pitfalls.

—Desi Arnaz (1917–1986)

During your doctoral study, the deeply experienced research community of your field—including your Ph.D. (co-)supervisor—is the author of the many papers you will be referring to; the author of various book chapters and books you would rely on throughout your doctoral investigations; and the anonymous reviewer of your prospective research works that will be submitted to conferences and journals. The intellectual achievements and standards of this seasoned research community can directly or indirectly influence your prospective Ph.D. research and intellectual journey, which we need to put in context. To do so, we need to estimate your intellectual level, the level of the seasoned research community in your field, and

T. M. Getu, *Pathways for Inspired Scientists and Engineers*,
https://doi.org/10.1007/978-3-032-06784-5_3

the gradient of the difference between the research community's scholarly caliber and yours, as a beginning researcher and Ph.D. student. These are all important to elucidate the nature, engagement, and capabilities of the *broader academic ecosystem* that you will informally join as a Ph.D. student and new researcher.

Empowered by its depth and breadth of scholarly experience and expertise, the seasoned research community of your field can essentially be adept at tracking and mastering the rapidly evolving state-of-the-art literature; sensing and tracking trendy research topics and themes; asking the right research questions; formulating and solving novel research problems; delivering original results of significance; spotting and identifying emerging and recurring scholarship opportunities; recognizing the disciplinary, interdisciplinary, multidisciplinary, and transdisciplinary research ecosystem; and networking and collaborating globally for win-win research discoveries and innovations. These impressive skills and capabilities constitute the *high scholarly caliber*—i.e., the high intellectual bar—that you need to qualify for, especially when targeting top-notch journals.

Contrary to the high scholarly caliber of the seasoned research community, you—as a Ph.D. student and new researcher—may lack significant scholarly experience and hands-on scholarly training in conducting original research independently. As a result, you may not be sufficiently experienced and skilled at tracking and mastering the rapidly evolving state-of-the-art literature; sensing and tracking trendy research topics and themes; asking the right, intellectually stimulating questions; formulating and solving innovative research problems; delivering novel results of significance and relevance; spotting and identifying emerging and recurring scholarship opportunities; recognizing the entire research ecosystem; and networking and collaborating globally for impactful scientific results.

From the above discussion, it is evident that there is a considerable gradient in how much seasoned researchers can track and master the rapidly evolving scientific literature in comparison with you; sense and track trendy research topics and themes in comparison with you; ask the right intellectually stimulating questions in comparison with you; formulate and solve novel research problems in comparison with you; deliver original results of significance in comparison with you; spot and identify emerging and recurring scholarship opportunities in comparison with you; recognize the entire research ecosystem in comparison with you; and network and collaborate globally to drive impactful innovations and discoveries. Consequently, you will face numerous chicken-and-egg problems that signify the inevitability of numerous *research-related Ph.D. pitfalls*, which are seasonally *specific Ph.D. pitfalls* that occur during the initial, middle, or final years of your Ph.D. study.[1]

As detailed in Chap. 2, you will be confronted by various Ph.D. pitfalls that are inevitable during the initial, middle, and final years of your Ph.D. study,

[1] During any of these periods, you will also come across numerous specific Ph.D. pitfalls as diverse as *writing and publication-related Ph.D. pitfalls, management-related Ph.D. pitfalls, and persona and motivation-related Ph.D. pitfalls*, which are the main themes of Chaps. 4, 5, and 6, respectively.

underscoring the inevitability of *not only specific but also generic Ph.D. pitfalls*.[2] Meanwhile, regarding the inevitability of several research-related Ph.D. pitfalls, this chapter provides a thorough examination of how and why they arise for the following reasons: first, to inform you in advance the likely research-related Ph.D. pitfalls you would face in your Ph.D. study; second, to prepare you in advance so that you will not get stuck by the research-related Ph.D. pitfalls that are inevitable during your Ph.D. study; third, to motivate you to turn proactively to the references cited by Chap. 3 and this book; fourth, to encourage you to discuss the research-related Ph.D. pitfalls with your network of peers; and last, to inspire you to develop *critical thinking, creative thinking, strategic thinking, probabilistic thinking, causal thinking, and independent thinking* capabilities, which you will dearly need throughout your Ph.D. study.

In accordance with the above purpose, this chapter presents detailed arguments on the various research-related Ph.D. pitfalls[3] inevitable during the initial and middle years of your Ph.D. study. Particularly, it details the numerous research-related Ph.D. pitfalls evident during the initial years of your Ph.D. study, such as getting to know the research ecosystem, identifying what is new to you and to others, too much state-of-the-art knowledge and advancements, managing immense and overwhelming literature, balancing between reading everything and being selective, identifying relevant and distracting literature, failing to identify constructive and destructive habits, feeling most innovations already exist, feeling incompetent, comprehending your Ph.D. (co-)supervisor's perspectives, understanding your study's limitations and delimitations, proposing research that could be a lifetime work, and getting new ideas. It also exposes the various research-related Ph.D. pitfalls unavoidable during the middle years of your Ph.D. study, such as forgetting the literature, failing to ask the right questions, failing to work backward from where you want to end, advancing from problems to ideas, advancing from ideas to problems to solutions, getting stuck, choosing easy research problems, comprehending and delivering originality, seeing what others have not seen, seeing the big picture and developing self-initiatives, tough comments of your Ph.D. (co-)supervisor, and setting a conceptual framework for your Ph.D. dissertation.

The remainder of this chapter is organized as follows. Section 3.2 details 13 research-related pitfalls inevitable during the initial years of your Ph.D. study. Section 3.3 follows with 12 research-related pitfalls inescapable during the middle years of your Ph.D. study. Finally, Sect. 3.4 ends this chapter with concluding remarks.

[2] Part II also discusses several generic Ph.D. pitfalls, as detailed in Chap. 7. If someone has already told you that there are many mountains to climb to earn your Ph.D., the specific and generic Ph.D. pitfalls are the ones you have to climb several mountains for.

[3] The to-be-discussed research-related pitfalls provide many lessons for a Ph.D. study and inspire research and thinking on the pitfalls detrimental to establishing a research career.

3.2 Research-Related Pitfalls During the Initial Years of a Ph.D.

If there is no struggle, there is no progress.

—Frederick Douglass (ca. February 1817 or 1818–1895)

Aiming for inspiration and pragmatic guidance, this section provides an in-depth discussion on the various research-related pitfalls inevitable during the initial years of your Ph.D. study. These likely research-related Ph.D. pitfalls are—but not limited to—listed below.

! Attention: Research-Related Pitfalls During the Initial Years of a Ph.D.

- *Getting to know the research ecosystem.*
- *Identifying what is new to you and to others.*
- *Too much state-of-the-art knowledge and advancements.*
- *Managing immense and overwhelming literature.*
- *Balancing between reading everything and being selective.*
- *Identifying relevant and distracting literature.*
- *Failing to identify constructive and destructive habits.*
- *Feeling most innovations already exist.*
- *Feeling incompetent.*
- *Comprehending your Ph.D. (co-)supervisor's perspectives.*
- *Understanding your study's limitations and delimitations.*
- *Proposing research that could be a lifetime work.*
- *Getting new ideas to work on.*

In what follows, we discuss each of the itemized research-related pitfalls.

3.2.1 Getting to Know the Research Ecosystem

The system, to a large extent, causes its own behavior! An outside event may may unleash that behavior, but the same outside event applied to a different system is likely to produce a different result.

—Donella H. Meadows (1941–2001)

Unless you are a Ph.D. student with significant prior research experience in the research field you would like to start your Ph.D. research on, it can be frustratingly difficult to realize where you are starting, especially concerning the hot topics and advancements of a trendy research field. To alleviate this difficulty and grasp where you are starting, you undoubtedly need to conduct a comprehensive survey on your research field's state-of-the-art literature that you must digest well. However, this would make you suffer from a substantial time constraint during the initial years of your Ph.D., when you are expected to conduct both coursework and research work concurrently.

Even if you succeed in diligently conducting a literature survey, you are susceptible to some illusion that may pop up in your head—with a high probability—due to the fundamental nature of the human brain. Because the human brain is a *causal biological machine* [1], it often thinks/perceives in terms of what it has gone through (i.e., its hands-on *accumulated neural experience*) and becomes conscious accordingly. Consequently, you may think in terms of what is inside your brain—i.e., *your intrinsic world*—and feel that you know well where you are starting; you have already grasped the full extent of your research field; and you are "a kind of done" with surveying the existing literature of your research field. Such a mindset can be due to the entitled research-related Ph.D. pitfall that can have multiple consequences.

While you are new to a research field you will work on for 3–5 years, senior researchers have already produced a wealth of contributions by reading, re-reading, questioning, experimenting, collaborating, persevering, imagining, and reimagining the field's then-state-of-the-art knowledge and advancements. As a result, a constraint will be implicitly imposed upon your level of emerging expertise, so you should climb all the *scientific ladders* to level up the accumulated state-of-the-art knowledge and innovations. Because senior researchers may have spent years, overcoming many challenges, to develop advanced expertise contributing to your research field's state-of-the-art knowledge and advancements, why wouldn't you thus take multiple years of persistent and passionate research work to develop a similar type and level of expertise?[4]

3.2.2 *Identifying What Is New to You and to Others*

> *Life is either a daring adventure or nothing at all.*
>
> *—Helen Keller (1880–1968)*

As a Ph.D. student and beginning researcher, you will learn and work to contribute through your publications while identifying what is new to you and to others. However, how much of a new thing is new? Is that until an anonymous reviewer reacts to it as a new one? Is that until a given research field becomes saturated? How and when? Since you would always think in terms of what is in your brain [1], as highlighted before, what is new to you—concerning "your intrinsic world"

[4] This harsh reality of being a new researcher usually entails a steep learning curve that shall be navigated with *purpose, persistence, and passion* for contribution.

embedded in your mind—may not necessarily be new to others, especially the leading experts of your research field. On the other hand, because you are not an expert researcher whose knowledge embodies the depth and breadth of your research field's advancements, you may find it hard, as a new Ph.D. student and researcher, to identify what is new to you and others.

Judging fiercely and objectively what is new to you and others requires you to develop substantial expertise. On the contrary, you need to discover several new things to develop significant expertise. Such a *chicken-and-egg problem* signifies this research-related Ph.D. pitfall, which can get compounded daily regarding the rapid advancements of a dynamic research field. For instance, what may feel surely new to you—regarding your knowledge encoded by your brain—could be old or even irrelevant to an expert. Accordingly, this pitfall can impact your decision concerning problem choice, research focus, and novelty.

3.2.3 Too Much State-of-the-Art Knowledge and Advancements

> *A road often traveled has heavy traffic and you take more time to reach the destination.*
>
> *—Ajeet Sharma*

According to the *data, information, knowledge, and wisdom pyramid (DIKW pyramid)* in [2, p. 55, Figure 2], knowledge amounts to organized information,[5] so is a research field's state-of-the-art knowledge. These days, it is easy to find too much information on a research field's state-of-the-art knowledge and advancements due to the many journal, magazine, conference, workshop, and special issue papers that are being frequently published by numerous not-for-profit professional associations and organizations. For-profit publishers also regularly publish abundant scientific literature to keep up with the trends of the twenty-first century's dynamic research enterprises. Consequently, each research field has produced abundant scientific literature of depth and breadth, whose availability is both an immense opportunity and an immense challenge to twenty-first-century researchers.

Although the availability of too much published literature can afford various opportunities for detailed understanding and innovation, it can cause anxiety, restlessness, a lack of focus, time constraints, confusion, and so forth to any new researcher. Hence, the challenges of too much published literature can outweigh its afforded opportunities, especially for a new Ph.D. researcher such as yourself,

[5] Per the DIKW pyramid, *data are abstracted elements; information is linked elements; knowledge is organized information*; and *wisdom is applied knowledge* [2, p. 55, Figure 2]. On the other hand, as per the *balance theory of wisdom* [3, Ch. 7], *wisdom is the application of successful creativity and intelligence—as intercede by values—to attain a common good while balancing interpersonal, intrapersonal, and extrapersonal interests over short and long terms to shape, adapt, and select environments* [3, p. 152].

making the availability of too much state-of-the-art knowledge and advancements one of the most unsettling research-related pitfalls during the initial years of a Ph.D. study. In light of this pitfall, you would likely be overwhelmed during the initial years of your Ph.D. study, provided that you undertake a Ph.D. project in a research field you have never researched. No wonder this could be the case for you as a new Ph.D. student, since you may not be skilled and experienced enough in managing and mastering the depth and breadth of literature pertaining to your research field.

3.2.4 Managing Immense and Overwhelming Literature

> *Many of us feel stress and get overwhelmed not because we're taking on too much, but because we're taking on too little of what really strengthens us.*
>
> —*Marcus Buckingham (1966-)*

Managing evolving knowledge and a vast amount of literature can be overwhelming for a beginning Ph.D. student who has never worked in a rapidly evolving research field. Whether the availability of huge literature is useful or not for the specific research problem you will formulate, it is probably overwhelming to you as a beginning Ph.D. student, who may be new to your Ph.D. research field. Although you may not know whether your research field's huge literature is instrumental until you creatively formulate a novel research problem to solve, your mind can be anxious and demotivated by the bulk of information embedded in the vast literature, which may be distracting. Hence, you can likely be confronted by the entitled research-related Ph.D. pitfall, which is inevitable during the initial years of your Ph.D. study.

Although you may attempt to manage your research field's immense and overwhelming literature to the best of your ability and knowledge, a research-related Ph.D. pitfall often arises because an effective management of such literature requires abundant experience in *effective literature search, efficient reading, and building the argument of discovery [4]; surveying, questioning, reading, recalling, and reviewing [5]; and assembling the data, organizing the data, and analyzing the claims [4].*

3.2.5 Balancing Between Reading Everything and Being Selective

> *You can't grow long-term if you can't eat short-term. Anybody can manage short. Anybody can manage long. Balancing those two things is what management is.*
>
> —*Jack Welch (1935–2020)*

As a new Ph.D. student preparing to navigate the vast and unknown academic landscape, you would be like *a baby brought to this world by its mom.* Because you will be entirely new for the twenty-first-century research enterprises and the intellectual landscape of your research field, your Ph.D. (co-)supervisor would typically expect you to conduct an intensive literature survey regularly. While surveying the possible enormous literature of your research field, you may feel as if you are swimming in an endless ocean, where you cannot navigate in the right direction, which makes you a step closer to your destination. Should you be selective enough to move on quickly, you may not have the advanced expertise that would inform your thoughtful decision on whether something is relevant. Consequently, you will face the entitled research-related Ph.D. pitfall, which can impact your research progress during the initial years of your Ph.D. study.

The entitled pitfall can hinder your research advancement during your initial Ph.D. years, since striking a perfect balance between reading everything and being selective may not be straightforward for you as a Ph.D. student. If you decide to be a voracious reader during the initial years of your Ph.D. study, you may gain much knowledge that comes at the price of spending considerable time. If you decide to be quite selective during your initial Ph.D. years, you may incur a lack of profound knowledge, which can steer your creativity and imagination, though you would have the advantage of saving ample time. Nevertheless, the time you would be able to save during the initial years of your Ph.D. study may put you at a disadvantage for the later years of your Ph.D. study, because the time you would save comes at the cost of shallow knowledge. The mentioned scenarios underscore the dilemmas of finding the *optimal balancing spot* that can be achieved via research experience, subject-matter expertise, and mature intellectual sensibility, which are the indispensable attributes that are hard to come by for a new Ph.D. student such as yourself.

3.2.6 Identifying Relevant and Distracting Literature

> *Science is built up with facts, as a house is with stones. But a collection of facts is no more science than a heap of stones is a house.*
>
> *—Henri Poincaré (1854–1912)*

According to the contemporary academic system, most Ph.D. students are expected to quickly and frequently publish journal and conference papers in order to finish their Ph.D. studies before running out of funding. As a result, there are many incremental papers knowingly and unknowingly published by the contemporary research enterprise, which has continued to advocate a *publish or perish game* in one way or another [6]. Even though many papers published by top-tier venues offer new ideas and opportunities for imagination and creativity, some may constitute *distracting literature* on top of several incremental papers driven by the current research enterprise. For this reason, you would be confronted by the entitled research-related pitfall during the initial years of your Ph.D. study, when you are expected to manage your coursework and research work concurrently, which should lead to publications in a short time.

When you are a new Ph.D. student working on a research field that you have never researched, identifying the relevant and distracting literature can indeed be a crippling research-related pitfall during the initial years of your Ph.D. study, especially concerning your endeavor of publishing as quickly as possible. This pitfall can be aggravated by the existence of your research field's many published papers, whose authors might be vying for *visibility and citation*.[6] It can also be aggravated by your lack of specialized expertise, research experience, and mature intellectual sensibility, which you will possess as you keep progressing with your Ph.D. research while surmounting its challenges.

3.2.7 *Failing to Identify Constructive and Destructive Habits*

> *Chains of habit are too light to be felt until they are too heavy to be broken.*
>
> —*Warren Buffett (1930-)*

Because your scholarly adventure toward your Ph.D. would be a long intellectual journey that needs a daily engagement, commitment, purpose, enthusiasm, independent thinking, and so forth, you may probably develop your habits intertwined with who you are—as a person and a researcher—and what you are engaged with. Each habit, bad or good, has its temporary reward that can convince you that you are doing the right thing. Since exhibiting a given habit is associated with your brain's reward system, your brain may feel comfortable staying in the comfort zone whenever you manifest any habit, even destructive ones. If the destructive ones continue to make your brain feel more at ease while staying in a comfort zone, the complacency of staying in the comfort zone will make your brain opt for destructive habits. Consequently, you will develop several destructive habits unsuspectingly, becoming your worst enemy throughout your Ph.D. study [7].

Considering you will be entirely new to a Ph.D. system and a research enterprise during the initial years of your Ph.D. study, you may unknowingly develop destructive habits while believing that you are engaging in constructive habits. Unfortunately, such development of destructive habits often offer an immediate reward for your brain, while it continually seeks your comfort zones adapted from your (under-)graduate study times. You may thus fail to identify constructive habits that you may perceive as destructive ones and vice versa. Consequently, you may fail

[6] The norms of the contemporary research enterprise can sometimes evoke pitfalls for next-generation researchers, like the pitfall of *identifying relevant and distracting literature*.

to identify *constructive habits*[7] and *destructive habits*,[8] which is a research-related pitfall that could be significantly detrimental, especially during the initial years of your Ph.D. study.

3.2.8 Feeling Most Innovations Already Exist

Science is the belief in the ignorance of experts.

—Richard P. Feynman (1918–1988)

During the initial years of your Ph.D. study, you are expected to regularly learn and discover the existing literature related to your Ph.D. project so that you develop sound subject-matter expertise, which will be a springboard for your upcoming research and contributions. Thus, your mind would be busy learning and thinking, making you wonder what to do for yourself. As you keep frequently discovering many published works, you would wonder more and more about the same thing. As this goes on, you may feel—with a high probability—that almost all important innovations of your research field already exist. As you keep getting this feeling, you may wind up getting depressed to the extent that you would somehow lose your self-confidence and belief in yourself, which make you pursue your Ph.D. study. This is the effect of the entitled research-related pitfall, which could be the most common feeling among new Ph.D. students in the initial years of their Ph.D. studies. As every researcher conceptualizes a new idea that one would investigate to publish a journal paper, many problems and themes of a given research field are unseen by the previous-generation researchers. However, figuring this out is difficult for a new Ph.D. student, who may not be an expert in their research field while suffering from the entitled research-related pitfall.

The entitled research-related pitfall, which you may believe cognitively during the initial years of your Ph.D. study, naturally translates to multiple chicken-and-egg problems. First, as you continue working creatively harder both in depth and breadth, you will realize via your emerging imagination and curiosity that numerous *big things* are awaiting discovery, given the astronomically many lines of research to investigate; however, you as a Ph.D. student may not possess the imagination and curiosity skills of an expert. Second, although you can deliver the foundational paper of your research field, which will surprise you with its endless "scholarly holes," delivering it requires advanced expertise and research insight, which you

[7] Constructive habits to be aware of: *making mistakes, communicating, giving credit where it is due, practicing finesse, and keeping an eye on the big picture* [7, p. 223–224]

[8] Destructive habits to be mindful of: *learned helplessness, taking ages to get nowhere, expressive behavior, isolation, and miscellaneous bad habits*, such as hiding from a Ph.D. (co-)supervisor, procrastination, bad time management, and failing to prioritize and put enough effort [7, p. 216–219]

may lack as a Ph.D. student. Third, you can dig deeper and broader with an expert mind to seize the gaps in the existing literature as opportunities; nevertheless, a new Ph.D. student may not be endowed with the expert insight that makes you spot them. Last, you can firmly believe like an expert that "science is the belief in the ignorance of experts,"[9] so your belief would lead to curiosity about promising investigations; however, you may not be an expert as a new Ph.D. student embarking on a new intellectual journey.

3.2.9 *Feeling Incompetent*

> *I was ordinary person who studied hard. There are no miracle people. It happens they get interested in this thing and they learn all this stuff, but they are just people.*
>
> *—Richard P. Feynman (1918–1988)*

As part of a comprehensive research survey you are expected to conduct during the initial years of your Ph.D. study, you may run into various "complicated" theorems, propositions, lemmas, corollaries, axioms, theories, etc. Since you may be entirely new to the research landscape of your Ph.D. research, you may be too busy grasping everything to develop significant expertise as quickly as possible. Despite your desire to move on, things may not be easy for your mind, which can keep you struggling to understand all those complicated innovations of the published authors. You may then suspect—with a high likelihood—that the published authors must have been "much smarter" than you, since they can disseminate "complicated works" that you may struggle to understand. As you continue struggling to understand, you may start being honest with your feelings, because our minds may need a minor distressing feeling to be pessimistic, so you must be incompetent (a.k.a. "stupid").[10] While this feeling is familiar among new Ph.D. students transitioning from a graduate study to a research study, it is a research-related Ph.D. pitfall that is inevitable during the initial years of your Ph.D. study.[11]

No wonder you may feel incompetent, since you will be a new Ph.D. student who will be entirely new to the global research enterprise and its Ph.D. system.

[9] When you believe that several things are either overlooked or ignored, you become curious to ask various questions that lead to promising investigations, affording multiple opportunities that can be seized via curiosity, persistent questioning, and even doubting.

[10] Very interestingly, great science may genuinely need *stupidity*! Here is a quote to ponder from a great researcher and leader: "The more comfortable we become with being stupid, the deeper we will wade into the unknown and the more likely we are to make big discoveries" [8].

[11] Feeling incompetent during the initial years of your Ph.D. study may cause feelings of *uncertainty and anxiety*. However, *these natural feelings do not signal incompetence but inexperience in conducting research, managing emotions and tasks, setting goals, and recognizing a learning experience [9].*

No wonder you may feel incompetent, since you may just got started reading the papers whose authors may have spent numerous hours on. No wonder you may feel incompetent, since you can hardly understand most published papers quickly, considering the efforts of authors who have intentionally spent numerous hours to satisfy the minimum novelty required for a publication. No wonder you may feel incompetent, since you must work persistently hard to arrive at the level of expertise attained by the leaders of your research field. No wonder you may feel incompetent, since you would mainly encounter tough journal papers whose reported perspectives were produced by the authors and expert anonymous reviewers, the *quality controllers*. No wonder you may feel incompetent, since you may come across that "tough paper" whose authors may have also been reading tons of literature to develop substantial expertise that led to the paper. No wonder you may feel incompetent, since several minds and accomplished researchers may have collaborated and thought through that "tough paper" before you discovered it as a publication. No wonder you may feel incompetent, since you need to read several papers to understand that "tough paper" easily. No wonder you may feel incompetent, since you may need to develop sufficient depth and breadth of expertise to find the scholarly substance of that "tough paper" much easier.[12]

3.2.10 Comprehending Your Ph.D. (Co-)Supervisor's Perspectives

Sometimes the best thing to do is to take a step back and get a little bit of a different perspective and re-evaluate things.

—Austin Aries (1978-)

It is broadly accepted that each of us is the product of our own experience, and so is your Ph.D. (co-)supervisor's various perspectives that embody depth and breadth of expertise. To this end, you may have difficulties comprehending the numerous perspectives of your Ph.D. (co-)supervisor, especially during the initial years of your Ph.D. study, for the following reasons. First, you may lack advanced knowledge and expertise in your research field. Second, you may be entirely new to the global research enterprise and the contemporary research landscape of your research field. Third, you may lack first-author publications in your research field or a related one. Last, you may be entirely fresh for the Ph.D. system of the region where your doctoral university is located. Consequently, comprehending the perspectives

[12] Although feeling incompetent may be common among starting Ph.D. students, no one has learned everything in the womb of one's mother. That is why even experienced researchers may feel anxious when they start "new kinds of projects on new topics" [9], and most published articles required a more extended time during their development and refinement, since "Rome is not built in a day."

of your Ph.D. (co-)supervisor[13] can be one of the major research-related pitfalls that you would face during the initial years of your Ph.D. study.[14]

Sometimes, being a Ph.D. student may feel like a horse pulling a carriage full of loads without a coachman, where a horse, a carriage, the loads, and a coachman are analogous to a Ph.D. student, a research field, the challenges of a research field, and a Ph.D. (co-)supervisor's directions, respectively. Hence, as a Ph.D. student, you may sometimes feel that you are pulling your research field forward while shouldering its challenges without necessarily having your Ph.D. (co-)supervisor's directions, especially when you fail to comprehend your Ph.D. (co-)supervisor's perspectives. Should you arrive at your desired destination (i.e., a novel solution) as efficiently as possible, you need an expert Ph.D. (co-)supervisor who would steer you in the right direction, regardless of your intensive effort and the research problem's toughness. To be perfectly steered toward the right directions, you should be able to comprehend the perspectives of your Ph.D. (co-)supervisor. If you have to succeed at this, however, you would need the mentioned qualities that you may lack as a Ph.D. student. Therefore, the entitled research-related pitfall causes several chicken-and-egg problems.

3.2.11 Understanding Your Study's Limitations and Delimitations

> *Remember, always, that everything you know, and everything everyone knows, is only a model. Get your model out there where it can be viewed. Invite others to challenge your assumptions and add their own.*
>
> *—Donella H. Meadows (1941–2001)*

When a researcher cannot control the factors that will or may affect one's study, they are called *limitations*; when a researcher can chiefly control those factors, they are termed *delimitations* [10, p. 114]. While delimitations are your choices on the boundaries of your study via justified assumptions that should be explicitly mentioned, limitations are the likely limits of your research field's state-of-the-art

---

[13] In neuroscience parlance, the *superposition* of the episodes of one's unique experience produces a unique *brain wiring (mind wiring)*. In this vein and probabilistically thinking, your comprehensive understanding of your Ph.D. (co-)supervisor's perspectives is likely achievable when you develop similar subsets of brain wiring—of your scholarly development in your research field—with your Ph.D. (co-)supervisor, whose brain wiring is exemplified by one's substantial research experience that gives rise to one's depth and breadth of expertise. This may be the reason why working with a renowned expert can be daunting, especially during the initial years of one's Ph.D. study.

[14] This could be true—with a high probability—provided that the underlying event is conditioned on other events pertaining to a lack of expertise and frustration due to failed experimentation.

knowledge and advancements, which are hardly under your control. For instance, your choice of the system model and its justifiable assumptions are the delimitations of your study, whereas all the factors regarding mathematical intractability and practical implausibility are its limitations.

Comprehensively understanding your research work's limitations requires depth and breadth of expertise in your research field and its state-of-the-art advancements. Contrarily, you, as a new Ph.D. student, may not have depth and breadth of expertise in your research field; you may endlessly swim in an ocean of detail-oriented literature, which can make you fail to grasp the big picture on the limitations of your study, especially during the initial years of your Ph.D. study. As a result, you may fail to understand the impacts of your study's limitations and allow some delimitations that are unjustified or of no significance. Therefore, you may face the entitled research-related pitfall, which can involve several chicken-and-egg problems.

3.2.12 Proposing Research That Could Be a Lifetime Work

If you do not understand your role in the problem, it is difficult to be part of the solution.

—David Peter Stroh (1950-)

In many North American universities offering Ph.D. programs, you are expected to define your Ph.D. project that outlines novel research problems, as part of the preparation and defense of your Ph.D. proposal, which is required for your Ph.D. candidacy. Possibly threatened by this high expectation, like many Ph.D. students, you may propose a Ph.D. project that could be a lifetime work. This can plausibly be due to lack of sufficient experience in your research field; shortage of scholarly experience in foreseeing research topics and estimating their required time to study; absence of background expertise to think through the depth and breadth of research questions that would demand answers; and inadequate project management skills [11, Ch. 3]. Because you will have limited time and funding for your entire Ph.D. study, you may likely face the entitled research-related Ph.D. pitfall during the initial years of your Ph.D. study.

Unless you have depth and breadth of expertise in your research field, which paves the way to formulating good research problems [12], your brain would not know the numerous intricacies and connections among the various concepts of your research field. This neuroscientific reality can make your brain overlook the likely challenges and open problems that can be roadblocks to the solutions of the research problems you would seek in your Ph.D. study. Accordingly, the failure to foresee the roadblocks will leave you with little to none *cognitive sensitivity* to the extent that you would wind up formulating *hard or infeasible research problems*, which you can hardly succeed in solving also due to the absence of depth and breadth of expertise, especially during the initial years of your Ph.D. study. Such chicken-and-egg problems can aggravate your inability to formulate concise and clear-cut research problems, which shall be formulated in a one bite at a time [7], compounding the effects of the entitled research-related Ph.D. pitfall.

3.2.13 Getting New Ideas to Work On

An idea is nothing more nor less than a new combination of old elements.

—James Webb Young (1886–1973)

After grasping your research field's historical background, relevant terminology, contemporary context, state-of-the-art developments and limitations, and foundational theories and concepts [5], you as a Ph.D. student are expected to start working on novel investigations that would lead to impactful results. During the initial years of your Ph.D. study, the success of such investigations would determine your fate as a prospective Ph.D. candidate. As a result, you may feel high pressure whenever you imagine or inquire about new ideas. This is the effect of the entitled research-related Ph.D. pitfall, which is one of the significant challenges of transitioning from a graduate student to a research student.

Getting new ideas can be as challenging to a new research student's brain as acrobatics is to a person who has never been a gymnast. As a gymnast has to do numerous acrobatics to be adept at it, a new research student has to engage with intensive independent thinking, conceptualization, and imagination to be skilled at generating new ideas. Since independent thinking, conceptualization, and imagination can hardly be successful without depth and breadth of expertise in your research field, you may struggle to get new ideas, which you wish to *explore*[15] for your prospective publications, as a new Ph.D. student investigating a research field you have never worked on. Therefore, the entitled research-related Ph.D. pitfall can lead to several chicken-and-egg problems and considerable uncertainties. Given this pitfall, one of the notable challenges of being a highly cited researcher is constantly getting new and impactful ideas.

We now move on to the research-related pitfalls inevitable during the middle years of your Ph.D. study.

[15] Any conceptualized idea can lead to promising results, additional investigations, or even "failures." However, several thinkers agree *there are no "failures" but wrong results*, which naturally afford many lessons and opportunities for directing an inquiry to a viable innovation path worth pursuing.

3.3 Research-Related Pitfalls During the Middle Years of a Ph.D.

Scientific research is not itself a science; it is still an art or craft.

—William H. George[16]

Depending on your progress and Ph.D. program's duration, the middle years of your Ph.D. study can last from the late second year to early/late fourth year of your Ph.D. study, when you become a Ph.D. candidate who has garnered *relatively significant* research experience.[17] During this period, you would face various research-related Ph.D. pitfalls that emanate from the inherently uncertain nature of your middle Ph.D. years and your research engagement during your initial Ph.D. years. As a result, you would come across the research-related Ph.D. pitfalls—but not limited to—that are itemized below.

! Attention: Research-Related Pitfalls During the Middle Years of a Ph.D.

- *Forgetting the literature.*
- *Failing to ask the right questions.*
- *Failing to work backward from where you want to end.*
- *Advancing from problems to ideas.*
- *Advancing from ideas to problems to solutions.*
- *Getting stuck.*
- *Choosing easy research problems.*
- *Comprehending and delivering originality.*
- *Seeing what others have not seen.*
- *Seeing the big picture and developing self-initiatives.*
- *Tough comments of your Ph.D. (co-)supervisor.*
- *Setting a conceptual framework for your Ph.D. dissertation.*

Despite the inevitability of the itemized pitfalls, you are expected to set a conceptual framework for your prospective Ph.D. dissertation; formulate novel research questions that you can successfully solve; generate original ideas that lead to your innovative solutions and results; undertake challenging research problems; handle

[16] William H. George was an English physicist who wrote a book entitled *The Scientist in Action: A Scientific Study of His Methods* [13].

[17] The word *significant* is relative. For a renowned scientist, significant research experience could be 15–20 years of research experience after earning a Ph.D. For a Ph.D. candidate, relatively significant research experience can be intensive research experience during one's initial Ph.D. study years.

your Ph.D. (co-)supervisor's challenging comments; neutralize your subjectivities and see what others have not seen; and so forth.

In the remainder, we discuss each of the itemized research-related pitfalls.

3.3.1 *Forgetting the Literature*

> *Originality is the fine art of remembering what you hear but forgetting where you heard it.*
>
> —*Laurence J. Peter (1919–1990)*

During the initial years of your Ph.D. study, you may need to read numerous research papers that you would deploy to write your Ph.D. proposal, which would inspire your future investigations toward addressing specific research problems whose solutions lead to your high-quality publications. However, because your envisioned intellectual path may lead to nowhere, you may decide to recap the evolving state-of-the-art research literature, which may make you realize that you forget most of what you read while being immersed in the intricacies of the problem you have been pursuing. This is one of the likely circumstances of the entitled research-related Ph.D. pitfall.

Due to the proliferation of information communication technologies and the ultra-competitive global economy of this twenty-first century, there are numerous advancements in almost all research fields. Consequently, you may happen to forget the literature anytime, like any Ph.D. student, as the volume of research you may need to deal with can monotonically increase all the time. Besides, since you as a Ph.D. student would be naturally absorbed in details (i.e., *small worlds*) that can make you unfocused on the big picture of your Ph.D. research, you may forget the research landscape of your research field.

3.3.2 *Failing to Ask the Right Questions*

> *The scientist is not a person who gives the right answers, he is one who asks the right questions.*
>
> —*Claude Lévi-Strauss (1908–2009)*

During the middle years of your Ph.D. study, you are typically expected to formulate challenging research problems—inspired by good research questions—that you can solve independently. During this pivotal research endeavor, on the one hand, you may end up formulating fundamental problems that are either mathematically intractable or known to take decades of persistent hard work, should you still succeed; on the other hand, you may wind up formulating seemingly challenging but simple problems that hardly lead to valuable insights. In these circumstances, your research would fail to deliver insightful results, irrespective of your hard work, for you would fail to ask the right questions beforehand.

According to Zina O'Leary, who is the author of a book entitled *The Essential Guide to Doing Research* [14], research questions serve to define an investigation, set boundaries, and provide direction [14, p. 28–29]. According to Roel Snieder and Ken Larner, who are the authors of a book entitled *The Art of Being a Scientist: A Guide for Graduate Students and their Mentors* [15], *questions always drive research* [15, Ch. 5]. The right questions thus derive the right research while serving as a foundation for the right problems. Meanwhile, asking the right questions often entails depth and breadth of expertise, good intellectual sensibility, experience in conceptualizing and asking good questions, and unbounded curiosity, which are all the successful seasoned researchers' traits you may lack during the middle years of your Ph.D. study. Accordingly, the entitled research-related Ph.D. pitfall translates to several chicken-and-egg problems.

3.3.3 Failing to Work Backward from Where You Want to End

In the middle of difficulty lies opportunity.

—Albert Einstein (1879–1955)

After internalizing your research field's state-of-the-art knowledge and advancements, you shall have a goal that will serve you as a GPS[18] destination goalpost. Once you set a goal, you can work backward from where you would like to end, rather than stumbling on random directions and being lost while expecting to find a goal [7, p. 112]. Although beginning from where you want to end sounds straightforward, you—as a Ph.D. student—may fail *to begin with the end in mind*, as is the case with *easier said than done* when it comes to new researchers. Unless you have developed the habit of beginning with the end in mind, you will likely fail to work backward from where you want to end. Even if you have developed this habit, you may fail to work backward from where you want to end for two main reasons.

First, working backward from where you want to end requires vivid and promising *mental creation* (i.e., *neuroscientific creation or first creation*), which hinges upon not only depth and breadth of expertise but also advanced imagination skills. Second, adequate cognitive experience—that has shaped your brain to start with the end in mind—determines your success in working backward to deliver a new *physical creation* (i.e., *second creation or realization*) starting from vivid and promising mental creation. Nonetheless, you—as a Ph.D. student—may lack depth and breadth of expertise, advanced imagination skills, and adequate cognitive experience of beginning with the end in mind. Accordingly, you may face several

[18] GPS: Global Positioning System.

chicken-and-egg problems that can make you fail to work backward from where you want to end,[19] aggravating the entitled research-related Ph.D. pitfall.

3.3.4 Advancing from Problems to Ideas

We all die. The goal isn't to live forever, the goal is to create something that will.

—*Chuck Palahniuk (1962-)*

While conducting sufficient background research on your Ph.D. topic, you would come across several research problems—whether important or not—outlined by the published authors as their open problems, research challenges, or research outlook. Armed with your acquired understanding, you may formulate new research problems whose novel solutions you would like to devise. Once you have a properly formulated problem, your next step is to have an idea of how to solve it. Because advancing from problems to ideas is *easier said than done*, this could be a troublesome research-related Ph.D. pitfall, which you will frequently face during the middle years of your Ph.D. study.[20]

Advancing from problems to ideas can be a *highly uncertain* intellectual process, since the idea useful for solving one important problem can be largely irrelevant for another one you would like to undertake. The inherent uncertainty in this intellectual process can be compounded by the fact that different novel problems will entail different *attack strategies and frameworks*, requiring different ideas on how to solve them. Thus, you must produce multiple novel ideas that pave the way for advancing from problems to ideas, which is hardly an easy undertaking for a Ph.D. student and a new researcher aiming to overcome the entitled research-related Ph.D. pitfall. To succeed in this endeavor, you require abundant hands-on experience in advancing from problems to ideas. As a Ph.D. student and new researcher, however, you can lack such ample scholarly experience. As a result, you may likely face various chicken-and-egg problems that can aggravate the entitled research-related Ph.D. pitfall.

---

[19] Since there are countless ways of traversing from point B to point A, there are astronomically many ways of working backward from where you want to end, requiring your intensive thinking on the most effective and promising way.

[20] For the same reason, you may also repeatedly encounter this research-related Ph.D. pitfall during the initial years of your Ph.D. study.

3.3.5 *Advancing from Ideas to Problems to Solutions*

By doubting we are led to question, by questioning we arrive at the truth.

—Peter Abelard (1079–1142)

Once you have grasped your research field's state-of-the-art advancements, you may aim to investigate a generic or specific idea that has made you optimistic about getting a paper out of it. In line with your idea, your next goal would be formulating a problem and providing its solution, which you are expected to assess and validate via simulations or computer experiments. In comparison with the existing state-of-the-art solutions, your validated solution must outperform or improve them. During such an intellectual endeavor, you will be confronted by the entitled research-related Ph.D. pitfall, since advancing from ideas to problems and from problems to solutions is a highly uncertain intellectual process.

Although your generated idea may have the potential for impact, the major challenge would be solving your formulated problem in a novel way and offering a solution that can enhance or outperform the existing solutions, if any. In your attempt of advancing from ideas to problems to solutions, you are thus likely to face two research-related intertwined Ph.D. pitfalls: *advancing from ideas to problems and advancing from problems to solutions*. To effectively advance from ideas to problems, you may likely face various chicken-and-egg problems emanating from your lack of significant scholarly experience in novelty-oriented, impact-oriented, and feasibility-oriented problem formulations. To effectively advance from problems to solutions via novel ideas on how to solve them, you may need to overcome two interwoven research-related Ph.D. pitfalls: *advancing from problems to ideas* and *advancing from ideas to solutions*. Advancing from problems to ideas is a research-related Ph.D. pitfall than can be aggravated by multiple chicken-and-egg problems, as highlighted in Sect. 3.3.4. Advancing from ideas to solutions is also a research-related Ph.D. pitfall that can be compounded by several chicken-and-egg problems, which originate from not only mathematical and numerical challenges but also a lack of depth and breadth of understanding around the formulated problems.[21]

[21] We assure you that there are always ways to advance from ideas to problems to solutions, regardless of any challenges, which you can surmount through persistent hard work, determination, frequent experimentation, depth and breadth of thinking, and flexibility.

3.3.6 Getting Stuck

I am not discouraged, because every wrong attempt discarded is another step forward.

—Thomas Edison (1847–1931)

It is widely accepted that conducting research is a never-ending cycle of thinking and experimentation to analyze, quantify, understand, and discover the underlying principles, limitations, characteristics, and tradeoffs of a given phenomenon or phenomena. As a part of this iterative intellectual process, which is usually uncertain, there will indeed be wrong results (not necessarily "failures"), confusions, doubts, frustrations, hesitations, and alike, especially when the intellectual process is independently executed by a new researcher—such as a Ph.D. student—who aims to perform novel research. Therefore, getting stuck is inevitable during the intellectual journey of any Ph.D. student such as yourself, irrespective of your research field, especially during the middle years of your Ph.D. study, when you are expected to deliver novel research results.[22]

According to Roel Snieder and Ken Larner, who are the authors of a book entitled *The Art of Being a Scientist: A Guide for Graduate Students and their Mentors* [15], you are not doing research if you are always having success [15, p. 69]. Getting stuck is an opportunity and a good sign that you are undertaking challenging research problems whose solutions you are attempting to devise. As a new Ph.D. student, nonetheless, you can stressfully get stuck due to improperly formulated problems, mathematically intractable problems, a lack of problem attack strategies and frameworks, a lack of ideas on solving problems, mathematical and numerical challenges, and a lack of depth and breadth of expertise. Apart from a lack of substantial scholarly experience as a Ph.D. student, any of these challenges can cause the entitled research-related Ph.D. pitfall, which can be crippling during the middle years of your Ph.D. study, since you are expected to deliver novel research results while also being aggravated by multiple chicken-and-egg problems.

3.3.7 Choosing Easy Research Problems

Every time we push personal development aside, we invite personal struggle into our lives.

—Hal Elrod (1979-)

During the challenging phase of your middle Ph.D. years, you may formulate easy research problems that you can quickly solve, so that you can get through your Ph.D. study as soon as possible. When your brain is in a frustration or gambling

[22] The entitled research-related Ph.D. pitfall can also happen during your initial Ph.D. years.

mode, choosing easy problems can feel like the right thing to do, since it may be a way to get a paper quickly out of your modest effort. However, it can be completely counterproductive for fulfilling the requirements of a Ph.D. (see Sect. 1.2), attaining the specific objectives of a Ph.D. (See Sect. 1.6.2), and establishing a prospective research career. Therefore, choosing easy research problems is a research-related Ph.D. pitfall that can trap you throughout the middle years of your Ph.D. study.

While your brain can understandably be in a state of possible frustration or gambling mode for choosing easy research problems during the middle years of your Ph.D. study, intentionally doing so can be counterproductive for your Ph.D. journey and future research career in many ways. First, solving easy problems may not help you demonstrate your ability to independently conduct original research of sufficient difficulty and scope since such problems can hardly lead you to top-tier journal, magazine, and conference publications. Second, resolving easy research problems may make you fail to attain the specific objectives of a Ph.D., because such problems can hardly develop you sufficiently to demonstrate your capability of lifelong self-learning, logical reasoning, and conducting independent and innovative research. Third, treating easy problems may likely fail you in establishing a prospective research career, since such problems can make you fail in accomplishing the requirements and specific objectives of a Ph.D., while failing you in publishing top-tier journal papers. As a result, choosing and solving several easy research problems cannot make you achieve a novel Ph.D. dissertation.[23]

3.3.8 Comprehending and Delivering Originality

> *Take the road less traveled. Get lost if you must, because only in losing yourself will you find your true path ....*
>
> —*Virginia Alison*

As a Ph.D. student, you are predominantly expected to deliver original research contributions that would be published by top-tier journals, which typically assess the *originality of your work* using expert anonymous reviewers. To successfully deliver originality, you must comprehend the notion of originality beforehand. On the contrary, you must first deliver original contributions to comprehend the notion

[23] When you are trapped by the entitled research-related Ph.D. pitfall, you may feel "accomplished" in addressing easy problems, so you tend to select only easy problems throughout your Ph.D. study, irreparably hurting your research career.

of originality, since your brain—as a causal biological machine—needs to draw on your scholarly experience and process of delivering originality. Accordingly, the entitled research-related Ph.D. pitfall would boil down to several chicken-and-egg problems you would face throughout the middle years of your Ph.D. study.

Originality can be grasped in many different ways, such as *complete originality*, *continued or modified originality*, *cross-pollination originality*, and *interdisciplinary, multidisciplinary, or transdisciplinary originality*. Due to a lack of significant scholarly experience in delivering originality, you may fail to comprehend complete originality, such as delivering an entirely new innovation or discovery for the first time; you may fail to comprehend continued or modified originality, such as demonstrating originality in modifying an existing idea, framework, or theory to address a new research problem; you may fail to comprehend cross-pollination originality, such as synthesizing novel algorithms for your research field by borrowing ideas or techniques from other research fields; and you may fail to comprehend (inter/multi/trans)disciplinary originality, such as creatively deploying (inter/multi/trans)disciplinary methodologies for addressing your novel research problem. When you may likely fail to comprehend complete originality, continued or modified originality, cross-pollination originality, and (inter/multi/trans)disciplinary originality, you may likely fail to deliver original works, asserting the impact of the entitled research-related Ph.D. pitfall.

3.3.9 Seeing What Others Have Not Seen

> *Discovery consists of seeing what everybody has seen, and thinking what nobody has thought.*
>
> *—Albert Szent-Gyorgyi (1893–1986)*

As underscored by the aforementioned quote, discovery in the context of your Ph.D. research would need two main ingredients: seeing what all your research field's published authors have seen and seeing what those researchers have not seen (or thought). Although anyone may advise you to see what others haven't seen so that you can make a scientific discovery, it is often the case that it is *easier said than done*. Since there could be several reasons why your research field's published authors did not see a given phenomenon, there could also be many reasons why you, as a Ph.D. student and new researcher, may not be able to see what others have not seen. This affirms the entitled research-related Ph.D. pitfall,[24] which you may likely face throughout the middle years of your Ph.D. study.

[24] We coined this Ph.D. pitfall in a figurative sense. However, it can also be named as *thinking what others have not thought*.

Like any deep learning-based artificial intelligence algorithm whose generalization power is proportional to the size and diversity of the dataset employed to train it, the human brain thinks in terms of the knowledge it has accumulated. Hence, as a Ph.D. student, you fundamentally think in terms of the pieces of knowledge stored in your brain. Given this fundamental limit of nature, there could be various reasons why seeing what others have not seen can be legitimately challenging. First, you may lack adequate expertise in your research field and fail to see several intertwined and interconnected scientific roads that can inspire new papers. Second, you may lack sufficient expertise in your related research fields, so you may not be able to seize opportunities available at the crossover of your Ph.D. research field and the related ones. Third, you may fail to spot a research gap (i.e., *scientific hole*) in the existing literature, so you may fail to pursue several lines of research and exploit many contribution rooms, which can all be afforded by a scientific hole. Last, you may not be endowed with remarkable synthetic and creative imagination skills, which would enable you to interconnect almost everything in entirely novel ways.[25]

3.3.10 Seeing the Big Picture and Developing Self-Initiatives

Research is what I'm doing when I don't know what I'm doing.

—Wernher von Braun (1912–1977)

Because you are normally expected to deliver several novel results during the middle years of your Ph.D. study, you would be endeavoring to address multiple research problems to the best of your ability. In doing so, you would probably be absorbed by the details of every research problem, since your brain would likely be laser focused on the small details of the research problems you may aspire to resolve. Consequently, you may likely fail to see the big picture, which can also be due to poor research planning or a lack of expertise in one's research field [7]. When you fail to see the big picture, you will likely fail to develop self-initiatives, affirming the inevitability of the entitled research-related Ph.D. pitfall.

As a Ph.D. student and new researcher, you may lack significant expertise in your research field. As you lack more and more expertise, the big picture that can arise from attacking specific problems gets darker and darker. Once you fail to see the big picture, you will likely fail in developing self-initiatives. Metaphorically, once you fail to see the big tree, you will fail to see its branches; once you fail to see the big forest, you will also fail to see its trees. When you fail to see the big picture, you will fail to see several details emerging from the big picture,

[25] Once you have developed significant expertise in your research field, seeing what others have not seen comes down to developing creative and synthetic imagination skills, which would offer you opportunities to engage in imagination inspired by any literature.

which are all opportunities for self-initiatives; similarly, when you fail to uncover various details by attacking multiple problems in your research field, you will fail to see the consequent big picture. Given this fact, you would need to successfully execute multiple novel investigations that can enlighten the big picture, whose comprehensive grasp would enable you to develop your self-initiatives on its details. Developing self-initiatives also requires the capability of advanced independent thinking, such as critical thinking, creative thinking, strategic thinking, probabilistic thinking, and causal thinking. Nevertheless, you may not successfully perform multiple novel investigations without depth of expertise, breadth of expertise, and advanced independent thinking capability, which you may lack as a Ph.D. student. Therefore, multiple chicken-and-egg problems can aggravate the entitled research-related Ph.D. pitfall, which could dwarf your research progress during the middle years of your Ph.D. study.

3.3.11 Tough Comments of Your Ph.D. (Co-)Supervisor

It is not the answer that enlightens, but the question.

—Eugene Ionesco (1909–1994)

Throughout your Ph.D. study, you are expected to successfully conduct original research under the guidance of an expert Ph.D. (co-)supervisor, who would work as *hand in glove* with you to pave the way for your success in your Ph.D. study. Because your Ph.D. (co-)supervisor would bring funding and expert guidance to the student-supervisor relationship, you are largely expected to return the investment by substantially contributing to one's research laboratory. Because your Ph.D. (co-)supervisor would take the commanding role of criticizing, challenging, complementing, and validating your entire research, you may thus get several tough expert comments throughout the middle years of your Ph.D. study, since you are expected to deliver several novel results, asserting the inevitability of the entitled research-related Ph.D. pitfall during the middle years of your Ph.D. study.

Your Ph.D. (co-)supervisor's comments are mere opportunities for improving your work and advancing the novelty of your research. However, your Ph.D. (co-)supervisor's tough expert comments can sometimes be intimidating and depressing. This could be because your Ph.D. (co-)supervisor's role guides, challenges, and inspires you throughout your Ph.D. study. Accordingly, you would likely get tough comments, which can be depressing and intimidating, should your Ph.D. (co-)supervisor question that your research needs magnificent push and rework to be of high quality; should your Ph.D. (co-)supervisor be convinced that the novelty of your research can hardly qualify for the minimum novelty required by top-tier journals; should your Ph.D. (co-)supervisor doubt that you are not working hard enough; should your Ph.D. (co-)supervisor believe that you are not productive enough; and should your Ph.D. (co-)supervisor conclude that you are not competitive enough. Should you fail to comply with what you are advised to proceed with, you may get tough comments frequently from your Ph.D. (co-)supervisor, because you have to open yourself and your mind to rise to the necessary challenges that are always opportunities.

3.3.12 Setting a Conceptual Framework for Your Ph.D. Dissertation

> *Research is something that everyone can do, and everyone ought to do. It is simply collecting information and thinking systematically about it.*
>
> *—Raewyn Connell (1944-)*

After getting some of your journal and conference papers published, your significant research activity during the middle years of your Ph.D. study would be setting a conceptual framework for your Ph.D. dissertation. Although your prospective and timely high-impact publications can serve as a foundation for your Ph.D. dissertation, setting a conceptual framework for your Ph.D. dissertation requires much more than the published papers, since it requires depth and breadth of emerging expertise in your research field. Accordingly, the entitled is the case of a *thing easier said than done* and hence a research-related Ph.D. pitfall.

According to Phil Agre, who is the author of a book entitled *Networking on the Network: A Guide to Professional Skills for PhD Students* [12], you would face several challenges when you write a talk about your Ph.D. research, including defining terms, explaining methodology, anticipating objections, making distinctions, identifying patterns, acknowledging unanswered questions, and outlining problems for future work [12]. According to the same author, here is a glimpse of setting a conceptual framework for a Ph.D. dissertation [12]:

> Of course, everyone tries to assign these points to a suitable place when preparing an outline. But many students find that the points just keep coming, as if a volcano were continually erupting in the middle of the thesis, causing a disorderly mass of troublesome junk to flow out toward the edges.

Overcoming the mentioned challenges has hardly been a straightforward intellectual endeavor, considering that setting a conceptual framework for your Ph.D. dissertation often entails several challenges. Particularly, you will be confronted by the challenges of conducting comprehensive literature survey; identifying early the trends in the state-of-the-art research works—and their methodologies—that are directly or indirectly related to your Ph.D. topic; outlining the overall challenges, opportunities, and prospects of your research field; thinking continually through the big picture while thinking through the depth and breadth of issues touching upon your Ph.D. research topic; and working backward strategically from where you want to end.

3.4 Concluding Remarks

Every accomplishment starts with the decision to try.
—John F. Kennedy (1917–1963)

The intellectually challenging nature of a Ph.D. study, compounded with the many uncertainties of a Ph.D. journey, gives rise to numerous Ph.D. pitfalls, which are generic or specific. Among the specific Ph.D. pitfalls that happen during the different periods of your Ph.D. study, this chapter detailed many research-related Ph.D. pitfalls that are inevitable during the initial and middle years of your Ph.D. study.

Specifically, we presented detailed arguments on the numerous research-related Ph.D. pitfalls evident during the initial years of your Ph.D. study, such as getting to know the research ecosystem, identifying what is new to you and to others, too much state-of-the-art knowledge and advancements, managing immense and overwhelming literature, balancing between reading everything and being selective, identifying relevant and distracting literature, failing to identify constructive and destructive habits, feeling most innovations already exist, feeling incompetent, comprehending your Ph.D. (co-)supervisor's perspectives, understanding your study's limitations and delimitations, proposing research that could be a lifetime work, and getting new ideas to work on. In addition, we presented detailed arguments on the various research-related Ph.D. pitfalls that are unavoidable during the middle years of your Ph.D. study, such as forgetting the literature, failing to ask the right questions, failing to work backward from where you want to end, advancing from problems to ideas, advancing from ideas to problems to solutions, getting stuck, choosing easy research problems, comprehending and delivering originality, seeing what others have not seen, seeing the big picture and developing self-initiatives, tough comments of your Ph.D. (co-)supervisor, and setting a conceptual framework for your Ph.D. dissertation.

Although we tried to discuss the most important research-related Ph.D. pitfalls inevitable during the initial and middle years of your Ph.D. study, our limited discussion—due to a space constraint—may not have covered the *full-blown depth and breadth* of issues interwoven with the discussed themes. For detailed insights and realities, we encourage you to reflect this chapter's references and discuss them with your network of peers.

References

1. Cicurel R, Nicolelis M (2015) The Relativistic Brain: How It Works and Why It Cannot Be Simulated by a Turing Machine. Kios Press, Sao Paulo
2. Kelleher JD, Tierney B (2018) Data Science, Ser. The MIT Press Essential Knowledge Series. MIT Press, Cambridge, MA

3. Sternberg RJ (2003) Wisdom, Intelligence, and Creativity Synthesized. Cambridge University Press, New York, NY
4. Machi LA, McEvoy BT (2016) The Literature Review: Six Steps to Success, 3rd edn. Corwin, Thousand Oaks, CA
5. Ridley D (2012) The Literature Review: A Step-by-Step Guide for Students, 2nd edn. SAGE, London
6. Geman D, Geman S (2016) Science in the age of selfies. Proc. Natl Acad Sci 113:9384–9387
7. Petre M, Rugg G (2010) The Unwritten Rules of PhD Research, 2nd edn. Open University Press, Berkshire
8. Schwartz MA (2008) The importance of stupidity in scientific research. J. Cell Sci. **121**(11):1771
9. Booth WC, Colomb GG, Williams JM, Bizup J, FitzGerald WT (2016) The Craft of Research, 4th edn. The University of Chicago Press, Chicago, IL
10. Mauch JE, Park N (2003) Guide to the Successful Thesis and Dissertation: A Handbook for Students and Faculty, 4th edn. Dekker, New York, NY
11. Finn JA (2005) Getting a PhD: An Action Plan to Help Manage Your Research, Your Supervisor and Your Project. Routledge, New York, NY
12. Agre P (2002) Networking on the Network: A Guide to Professional Skills for PhD Students, 2nd edn. University of California–Los Angeles, Los Angeles, CA
13. George WH (1936) The Scientist in Action: A Scientific Study of His Methods. Williams and Norgate, London
14. O'Leary Z (2004) The Essential Guide to Doing Research. SAGE, London
15. Snieder R, Larner K (2009) The Art of Being a Scientist: A Guide for Graduate Students and their Mentors. Cambridge University Press, New York, NY

Chapter 4
Writing and Publication-Related Pitfalls of a Ph.D.

We must all suffer one of two things: the pain of discipline or the pain of regret and disappointment.

—Jim Rohn (1930–2009)

Abstract Publication-related Ph.D. pitfalls are due to a lack of publishing experience in conjunction with the inevitable research-related and writing-related Ph.D. pitfalls. In causing publication-related Ph.D. pitfalls, writing-related Ph.D. pitfalls can be as significant as research-related Ph.D. pitfalls since many Ph.D. students are new writers with limited authorship experience. Accordingly, this chapter discusses in detail the inevitable writing and publication-related Ph.D. pitfalls so that you will not get stuck by them throughout your Ph.D. study.

Keywords Writing-related Ph.D. pitfalls · Publication-related Ph.D. pitfalls · Writing and publication-related Ph.D. pitfalls

4.1 Introduction

If you want something you've never had, you must be willing to do something you've never done.

—Thomas Jefferson (1743–1826)

As a Ph.D. student and new researcher, you have little to no experience in professional publishing and writing. Regardless, your prospective conference and journal papers will be reviewed and critiqued by not only your Ph.D. (co-)supervisor but also the sufficiently experienced research community of your field, which has been setting intellectual and academic standards for your field. This *cold truth* of a Ph.D. life and academia will directly or indirectly evoke several writing and publication-related Ph.D. pitfalls due to your lack of experience in publishing, many inevitable research-related Ph.D. pitfalls, and various inevitable writing-related Ph.D. pitfalls, as argued below.

T. M. Getu, *Pathways for Inspired Scientists and Engineers*,
https://doi.org/10.1007/978-3-032-06784-5_4

The broader seasoned research community of your field—who will evaluate your prospective conference and journal papers—is substantially experienced in publishing new scientific results and driving science forward. By contrast, you have little to no publishing experience, especially in any prestigious venue, which can likely hamper your ability and strategy to manage the many publication-related uncertainties. As a result, due to the significant difference between the seasoned research community's ample publication experience/achievements and your relatively insignificant publication experience/achievements, you will undoubtedly face numerous publication-related Ph.D. pitfalls throughout your Ph.D. study.

Contrary to the remarkably high scholarly caliber of the seasoned research community, you—as a Ph.D. student and new researcher—may lack significant scholarly experience and hands-on scholarly training in conducting original research independently. Consequently, you may not be sufficiently experienced and proficient at sensing and tracking trendy research topics and themes; tracking and mastering the rapidly evolving state-of-the-art literature; asking the proper intellectually stimulating questions; formulating and solving innovative research problems; delivering original results of significance and relevance; recognizing the broader disciplinary, interdisciplinary, multidisciplinary, and transdisciplinary research ecosystem; spotting and identifying emerging and recurring scholarship opportunities; and networking and collaborating globally for impactful innovations and discoveries. All these inevitable research-related Ph.D. pitfalls make many publication-related Ph.D. pitfalls inevitable throughout your Ph.D. study.

As a Ph.D. student and new researcher, you have little to no professional writing experience, which is compounded by your lack of adequate training in advanced academic, non-academic, and technical writing. In this vein, you may struggle with paper writing throughout your Ph.D. study for several reasons. First, you may fail to structure your prospective conference and journal papers properly. Second, you may lack a sufficient writing experience to be concise and fluid in your language. Third, you may not be able to properly handle your prose in any of the papers you will write. Fourth, you may have difficulties compressing your voluminous work into a short conference/journal paper. Fifth, you may struggle to write a compelling story out of each of your prospective research works. Last, and importantly, you may fail to see the various layers of editing that are indispensable for producing an excellent paper. These are some of the inevitable writing-related Ph.D. pitfalls, which often give rise to many publication-related Ph.D. pitfalls that will be inevitable throughout your Ph.D. study. In causing publication-related Ph.D. pitfalls, writing-related Ph.D. pitfalls can be as considerable as research-related Ph.D. pitfalls because writing is a craft wherein no one has become a master, which arguably takes one's life to demonstrate [1–9], and almost all Ph.D. students are new writers with little to no authorship experience.

The above arguments justify the need to offer detailed discussions on the writing and publication-related Ph.D. pitfalls that are inevitable throughout your Ph.D. study. Accordingly, this chapter devotes considerable attention to how and why they arise, for the following reasons: first, to inform you in advance the probable writing and publication-related Ph.D. pitfalls you would face in your Ph.D. study; second, to prepare you in advance so that you won't get stuck by the writing and publication-related Ph.D. pitfalls inevitable during your Ph.D. study; third, to motivate you to turn proactively to the references cited by Chap. 4 and this book; fourth, to encourage you to discuss the writing and publication-related Ph.D. pitfalls with your network of peers; and last, to inspire you to develop *critical thinking, creative thinking, strategic thinking, probabilistic thinking, causal thinking, and independent thinking* capabilities, which you will need throughout your Ph.D. study.

For the reasons mentioned above, this chapter presents detailed and crucial perspectives on the various writing and publication-related Ph.D. pitfalls that are inevitable during the initial and middle years of your Ph.D. study. Particularly, it details the many writing and publication-related Ph.D. pitfalls that are likely during the initial years of your Ph.D. study, such as boss pressure for a publication; neutralizing your subjectivities; compressing a voluminous work into a short paper; failing to see and manage the different layers of editing; realizing how much evidence, reasoning, and acknowledgment are enough; a lack of experience in writing a paper; and a lack of experience in publishing a paper. It also details several writing and publication-related Ph.D. pitfalls that are likely during the middle years of your Ph.D. study, such as handling considerable boss pressure for a publication, a lack of advanced experience in telling a story, a lack of advanced experience in structuring a paper, a lack of advanced experience in writing a paper, and a lack of advanced experience in publishing a paper.

The remainder of this chapter is organized as follows. Section 4.2 discusses the writing and publication-related Ph.D. pitfalls inevitable during the middle years of your Ph.D. study. Section 4.3 exposes the writing and publication-related Ph.D. pitfalls inevitable during the middle years of your Ph.D. study. Finally, Sect. 4.4 ends this chapter with concluding remarks.

4.2 Writing and Publication-Related Pitfalls During the Initial Years of a Ph.D.

Embrace the uncertainty. Some of the most beautiful chapters in our lives won't have a title until much later.

—Bob Goff (1959-)

During the initial years of your Ph.D. study, you are expected to publish original papers—while concurrently managing both coursework and research work—which necessitates several skill sets for conducting research and publishing papers. On the contrary, you may have little to no experience in paper writing and publishing. Consequently, you would face the writing and publication-related Ph.D. pitfalls—but not limited to—itemized below.

! Attention: Writing and Publication-Related Pitfalls During the Initial Years of a Ph.D.

- *Boss pressure for a publication.*
- *Neutralizing your subjectivities.*
- *Compressing a voluminous work into a short paper.*
- *Failing to see and manage the different layers of editing.*
- *Realizing how much evidence, reasoning, and acknowledgment are enough.*
- *A lack of experience in writing a paper.*
- *A lack of experience in publishing a paper.*

We now proceed to discuss each of the itemized pitfalls one by one.

4.2.1 Boss Pressure for a Publication

> *Everything negative – pressure, challenges – is all an opportunity for me to rise.*
>
> —*Kobe Bryant (1978–2020)*

During the initial years of your Ph.D. study, you are expected to vividly demonstrate your scholarly worth that can qualify you for a Ph.D. candidacy while simultaneously managing your coursework and research work. As managing two things at once is hardly easy for a new researcher and Ph.D. student, this phase of your Ph.D. study can thoroughly test your intelligence, inherent motivation, and stamina. Your stamina would also be tested by your Ph.D. (co-)supervisor's *pressure for a publication*, which you may likely receive, provided that you seem to fail in demonstrating a promising research engagement and scholarly independence. You may wonder why it is so from the perspective of both yourself and your Ph.D. (co-)supervisor since everything shall be seen from the perspective of yourself and your employer [10].

As everything has a reason, you as a Ph.D. student shall be aware why you are expected to publish quickly to demonstrate your scholarly worth since being part of the solution starts with having a complete understanding—as part of your wisdom—of *the reality that can bite*. To begin, your Ph.D. (co-)supervisor would offer you a Ph.D. position when one has an adequate research grant that can cover your overall living expenses and tuition fees, if any. Such a research grant is often tied to a contract with a funding organization requiring your Ph.D. (co-)supervisor to report one's research progress regularly. Since you would be hired to conduct research under the auspice of your Ph.D. (co-)supervisor, you would be expected to be one's research assistant who can deliver research progress reports in the form of publications, possibly in top-tier journals and conferences. Because the granting funding agency cannot wait for your Ph.D. (co-)supervisor for years to submit one's timely research progress report, one cannot thus wait for years until you produce your first publication. Since publishing is usually an uncertain intellectual undertaking with its inherent risk, especially when executed by a new researcher, hiring you comes with an ingrained risk for your Ph.D. (co-)supervisor—in terms of one's valuable time, investment, and future research trajectory—who has to make sure that you are the right fit for one's funded research project. As a result, you may likely be pushed by your Ph.D. (co-)supervisor, who may feel doing the right thing in pressuring you for a publication.

Let us now move to the main reasons from the perspective of yourself. Because your brain is a *causal biological machine* [11]], it thinks and perceives in terms of what it has consciously gone through, such as the limited research pressure you may be levied during your undergraduate and graduate studies. Regarding this experience, your Ph.D. (co-)supervisor's minor demands may feel like tremendous pressure for a publication since you may not have felt the scholarly pressure you would be subjected to during the initial years of your Ph.D. study. For the same reason, you may easily feel immense pressure for a publication out of your Ph.D. (co-)supervisor's possible legitimate feedback that you are not working hard enough in terms of your creativity and productivity one may have objectively assessed; legit belief that you may not be as creative and productive as other Ph.D. students who also receive similar amount of Ph.D. scholarship; a logical supervisory practice that one has to test your capability of working under pressure or deadlines; firm belief that you are a type of researcher who would need adequate pressure to be productive enough; and conviction you have to rapidly *rise to the challenges of working under pressure or deadlines*, as it would be the case for many research projects.

4.2.2 Neutralizing Your Subjectivities

> *Complete objectivity is not an option. We are all subjective about the way we respond to 'what is,' whether it's the people we encounter, the circumstances in our lives, or ourselves. What we can do is reduce our subjectivity – what I call 'I see, therefore it is'.*
>
> —*Elizabeth Thornton*

After tirelessly researching for multiple months during the initial years of your Ph.D. study, you may complete a "great paper," which you may feel like a perfect work that a top-notch publication venue should immediately accept. As you would be emotionally involved in your work for a long time, your mind would likely find ways to believe that whatever you do is essential and perfect. Although this may seem a fitting subjective feeling for hard work, this human sentiment is precisely what would make you biased toward your work while filling you up with *subjective feelings (or emotions)* that can only orient you to judge or act subjectively.

Neutralizing such subjectivities is a writing and publication-related Ph.D. pitfall that can be significantly detrimental to your progress in research and publication, especially during the initial years of your Ph.D. study.

We all are *neuroscientifically subjective creatures*, and so are you, who also have your unique likes/dislikes, emotions, feelings, preferences, etc. [12]. As a neuroscientifically subjective creature and new Ph.D. student, you may be attached to your ideas, referred papers, approaches, research methodologies, frameworks, simulations or computer experiments, results and their interpretations, investigations, writing, writing style, etc. However, everyone's work has limitations and weaknesses from a fiercely objective standpoint, like your work, which can have weaknesses in its research design, choice of framework, research methodology, depth and breadth of reasoning, review of related works, arguments, corroborating your claims, mathematical proofs, design of experiment or simulations, performance comparison, writing, writing style, and recommendations. This is why any paper could be rejected concerning infinitely deep and/or wide perspectives.

Your subjectivities can disengage your *analytical mind* while continuously engaging your *emotional mind* to the extent that your work would be stripped of the essential critical thinking that commonly shapes many scientific papers. As a result, your subjectivities—unless neutralized—can impact your thinking, reasoning, argument, writing, criticality, experiments, and alike, hurting your progress in successful publishing and research. Although the human brain can be rigorously trained to be a fiercely objective biological machine, your level of thinking—as a Ph.D. student and new researcher—may not be fiercely objective to neutralize your subjectivities sufficiently. Specifically, neutralizing your subjectivities requires the development of an objective mindset that hinges on advanced critical and independent thinking capabilities, sufficient scholarly experience in conscious objective thinking, and considerable intellectual experience in objectively critiquing others' thinking. Nevertheless, you may lack all these attributes as a Ph.D. student and new researcher. Therefore, the entitled writing and publication-related Ph.D. pitfall boils down to many irreducible chick-and-egg problems.

4.2.3 Compressing a Voluminous Work into a Short Paper

Not that the story need be long, but it will take a long while to make it short.

—Henry David Thoreau (1817–1862)

During the initial years of your Ph.D. study, you may be working hard on your first important research problem for months by surveying several papers; formulating a new problem and proposing a new algorithm; proving demanding theorems that quantify your new algorithm's performance; carrying out extensive computer experiments and/or simulations; writing the entire procedure and results into a detail-oriented voluminous draft; etc. In light of this promising endeavor, your Ph.D.

(co-)supervisor may have neither a plan to pay extra page charges nor a conviction that your results should be reported in a long paper, forcing you to compress your written voluminous draft.

Since you may have to compress your voluminous draft into a succinct paper of a few pages, in line with the strict page limits of your research field's journal or conference, you have to *concisely* write the most important and unavoidable aspects of your work so that your compressed paper demonstrates considerable *clarity and simplicity* for your Ph.D. (co-)supervisor and prospective anonymous reviewers. To this end, your Ph.D. (co-)supervisor may provide you with many comments—for guidance and quality control—during your draft paper's multiple iterations of editing, which can cause frustration due to the additional hard work required to disseminate your research results. This is likely how you would face the entitled writing and publication-related Ph.D. pitfall, which can also confront any new Ph.D. student during the initial years of one's Ph.D. study.

As a new researcher preparing a paper for a conference/journal submission, you would join a new ecosystem whose members, players, rules, and dynamics are hardly clear to you. No wonder your Ph.D. (co-)supervisor may give you many comments regarding your first draft paper, because a Ph.D. (co-)supervisor is a successful researcher who completed many research projects before earning one's academic position; fulfilling the research and publication standard of an expert Ph.D. (co-)supervisor requires gigantic effort to be undertaken persistently, especially for a new Ph.D. student; a Ph.D. (co-)supervisor often acts as an internal and constructive reviewer as a testament of one's research quality; new writers such as yourself are often heard struggling to quickly extract a paper's take-home message, which they got to yet learn through practice; filtering and producing a *paper's juice* can be complex for new researchers such as yourself, since there would be significant inherent intricacies in a concluded research work; and a Ph.D. (co-)supervisor has to be convinced by the quality of your research and writing to approve your edited paper for a submission. Furthermore, successfully compressing a voluminous work into a short paper hinges on considerable editing and re-writing experience in paper compression, since it usually is different from paper to paper. However, as a Ph.D. student and new researcher, you may not have garnered considerable editing and re-writing experience.[1] Therefore, the entitled writing and publication-related Ph.D. pitfall boils down to many irreducible chicken-and-egg problems.

[1] Pay attention to how your field's leaders wrote their papers, which usually offer crucial lessons in neat organization, elegant prose, catchy flow, consistency, conciseness, objectivity, and sound reasoning.

4.2.4 *Failing to See and Manage the Different Layers of Editing*

The first draft is black and white. Editing gives the story color.

—Emma Hill

After working persistently hard for months during the initial years of your Ph.D. study, you may produce your first draft paper that can make you feel accomplished to conclude your first work during the early stage of your Ph.D. study. While wishing for validation and appreciation for your months of hard work, you may get numerous *red-line comments* from your Ph.D. (co-)supervisor. Although this may seem surprising or frustrating for a new researcher, senior researchers can even receive several editing-related comments from their professors, let alone a Ph.D. student who may likely lack adequate paper writing, editing, and publishing experience. This is because of the various layers of editing, which you may fail to see and manage during the initial years of your Ph.D. study. Hence, you may encounter the entitled writing and publication-related Ph.D. pitfall during this time.

Even if your paper is high-quality, you may get it rejected for average-quality editing work while failing to see and manage the different layers of editing. Since you, as a Ph.D. student, would often be working on the intricacies of your specific research problem, you often end up being embedded in your *small world of details*. When this reality conspires with your limited or no experience in paper writing, editing, and publishing, you would likely fail to see *the big picture* of high-quality writing, making you fail in managing the many layers of editing. Nonetheless, high-quality writing is always the result of several layers of editing, exemplifying *writing as a process of rewriting* [13, Ch. 6]. In this vein, as a Ph.D. student and new researcher, you are expected to manage the multiple layers of editing for *language and grammar, style, consistency, clarity, readability and simplicity, conciseness, rigor, accuracy, relevance, logicalness, depth, breadth, significance, and completeness* of a paper.

As a Ph.D. student and new researcher, you may not have adequate writing, editing, publishing, and scholarly experience to manage the language and grammar of your paper such that its entire grammar and usage is correct along with unambiguous wording; the style of your paper such that it is written with the widely accepted style of English writing; the consistency of your paper such that its style, symbols, wording, punctuation, abbreviations, and forms are consistent throughout; the clarity of your paper such that it is crystal clear on a line-by-line basis; the readability and simplicity of your paper such that it is readable and simple all over; the conciseness of your paper such that it is consistent throughout; the rigor of your paper such that its theorems, lemmas, corollaries, and propositions have rigorous proofs; the accuracy of your paper such that it is accurate line by line; the relevance of your paper such that its importance is substantiated everywhere; the logicalness of your paper such that its arguments are logical; the depth of your paper such that it features a sound depth of research; the breadth of your paper such that it illuminates a reasonable breadth of research; the significance of your paper such that it manifests consequential research; and the completeness of your paper such that its reported work is exhaustive. Therefore, the entitled writing and publication-related Ph.D. pitfall comes down to many irreducible chicken-and-egg problems.

4.2.5 Realizing How Much Evidence, Reasoning, and Acknowledgment Are Enough

There are no facts, only interpretations.

—Nietzsche (1844–1900)

As a hard-working and creative Ph.D. student, you may obtain some interesting results that you would present as evidence for the claim you wish to make in your conference/journal paper. Because you may probably spend considerable time and energy to produce those results, you may believe that you have already obtained the "bedrock evidence," so you can make a substantial claim in your conference/journal paper. However, your evidence may not be significant enough for an expert; you are expected to support your obtained evidence with reasons; and you are also expected to acknowledge other points of view. Hence, among the multifaceted issues of getting published by top-tier conferences and journals, you may likely face the entitled writing and publication-related Ph.D. pitfall during the initial years of your Ph.D. study.[2]

When you are new to the research enterprise, you need some time to learn the kinds of evidence accepted and rejected by the anonymous reviewers of your research field [14, Ch. 9]. Since expert reviewers are the ones who have seen several pieces of evidence in their research careers, seemingly significant evidence may wind up being insignificant to their eyes and minds. When you argue to claim anything, you are normally expected to support it with clear-cut evidence and a reason. Because a reason and evidence are hardly enough, you are usually encouraged to strengthen your argument with more reasons, which would, in turn, be supported by additional reasons and evidence [14, Ch. 9]. Aside from your reasons and evidence, you are also required to position your claims as contributions to the ongoing conversation by your readers while acknowledging other points of view to establish a cooperative relationship with your readers [14, Ch. 9].

Because estimating the required amount of evidence is intertwined with the novelty, quality, and criticality needed to get published by top-tier journals and conferences, it requires substantial research and publication experience, which you may lack as a Ph.D. student and new researcher. Since the required amount of reasoning is often correlated with the required amount of evidence, estimating the required amount of reasoning also requires considerable research and publication

[2] You could also face the entitled Ph.D. pitfall during the middle years of your Ph.D. study.

experience, which you may lack as a Ph.D. student and new researcher. Because estimating the required number of acknowledgments depends on your ability to foresee your readers' perspectives and questions, which you should address in advance, you may not be adept at this estimation since you may lack advanced imagination skills and expertise. On the same grounds, you may find it difficult to determine *the sweet spot* on the number of required acknowledgments, as acknowledging many and a few points of view makes you appear distracting and ignorant. Therefore, the entitled writing and publication-related Ph.D. pitfall comes down to several irreducible chicken-and-egg problems.

4.2.6 A Lack of Experience in Writing a Paper

Almost all good writing begins with terrible first efforts. You need to start somewhere.

—Anne Lamott (1954-)

The four significant constraints to consider in scientific writing are its *audience, format, mechanics, and surrounding politics* [15]. You also need to control your *writing style* by controlling your paper's *structure, language, and illustration* [15] while trying to handle its content correctly. Trying to handle all the mentioned attributes as a Ph.D. student and new writer, however, you will face several writing and publication-related Ph.D. pitfalls, which come down to numerous irreducible chicken-and-egg problems. This is because managing all the mentioned attributes of scientific writing requires considerable experience; nonetheless, you—as a Ph.D. student and new writer—may likely lack writing experience. Therefore, some of the writing-related Ph.D. pitfalls emanating from a lack of experience in writing a paper are—but not limited to—listed below.

! Attention: Writing-Related Ph.D. Pitfalls Due to a Lack of Experience in Writing a Paper

- *A lack of experience in structuring a paper.*
- *A lack of experience in handling the language of a paper.*
- *A lack of experience in handling the content of a paper.*
- *A lack of experience in handling the illustration of a paper.*

In what follows, the itemized Ph.D. pitfalls are discussed one by one.

4.2.6.1 A Lack of Experience in Structuring a Paper

Organization is what you do before you do it, so when you do it, it's not all messed up.

—Winnie the Pooh, A.A. Milne

For a new researcher who may lack adequate paper writing experience and knowledge, your brain *may not be sufficiently conscious* of the rules and practices for organizing the beginnings, middles, and endings of your paper, providing proper transition, depth, and emphasis to your paper [15].

Concerning the beginning of your paper, you may not easily create a fitting paper title that can correctly identify your field of research while distinguishing your paper from all other papers of your research field; compose a descriptive summary (also a descriptive abstract) and an informative summary (also an executive summary) that present the kind of information your paper will contain and the actual results of your work, respectively; and write an excellent introduction that states what your work is all about, why your work is important, what one needs to understand your work, and how your work will be presented [15]. Regarding the middle of your paper, you may not easily handle the logical strategies in scientific writing—i.e., chronological, spatial, flow, or cause-effect strategies—that should be conveyed to your readers via the headings and subheadings you chose [15]. As for the ending of your paper, you may not produce a well-written conclusion that provides readers an analysis of the most important results of your work and its research outlook and a well-written appendix that offers additional/detailed information to the secondary audiences of your paper [15].

Even if you may succeed in properly organizing your paper, it may fail to inform your readers since you may not consciously make strong transitions between the details, present the details at the proper depth, and properly emphasize the details (e.g., using repetition, wording, illustration, and placement) [15].

4.2.6.2 A Lack of Experience in Handling the Language of a Paper

I have only made this letter longer because I have not had the time to make it shorter.

—Blaise Pascal (1623–1662)

As a new researcher who may lack adequate paper writing experience and knowledge, your brain may not be *sufficiently conscious* of handling the language aspects of your paper, such as *being precise, clear, forthright, and familiar* [15]. Precision is the most important goal of scientific writing since failing to communicate what you did in writing amounts to changing your work [15]. As to clarity in scientific writing, it means avoiding things that you do not mean [15]. As for forthrightness, a writer should be sincere and straightforward by controlling one's tone and selecting strong nouns and verbs [15]. Concerning familiarity in scientific writing, the writer is always responsible for bridging the language gap, not the reader [15].

However, as a Ph.D. student and new writer, you can fail in being precise by failing to choose the right word and the right level of detail; in being clear by failing to avoid *needless complexity* (e.g., needlessly complex words, phrases, and sentences) and *ambiguity* (e.g., ambiguities in word choice, syntax, pronouns, and punctuation); in controlling your tone by failing to avoid pretentious words, arrogant phrases, and silliness and failing to use strong nouns and verbs; and in being familiar—while embedded in a disciplinary jargon—by failing to avoid unfamiliar terms, define unfamiliar terms, and incorporate analogies and examples [15].

4.2.6.3 A Lack of Experience in Handling the Content of a Paper

> *Examine every word you put on paper. You'll find a surprising number that don't serve any purpose.*
>
> *—William Zinsser (1922–2015)*

Because you are a new researcher who may lack adequate paper writing experience and knowledge, your brain may not be *sufficiently conscious* of handling the various aspects concerning the content of your paper. Considering the *content guidelines* in [16, p. 147–154], you—as a Ph.D. student and new writer—can inadvertently make many mistakes in handling the content of your paper, even if you may succeed in conducting novel research. First, you may fail to state your problem and start strong clearly. Second, you may fail to make readers interested by not being able to connect your research theme to their concerns. Third, you may write an incomplete literature review that lacks focus and balance. Fourth, you may fail to demonstrate how your work builds on relevant past works. Fifth, you may fail to explain clearly what your results imply. Sixth, you may conclude beyond your data, negligently overreaching. Seventh, you may fail to clarify the limitations of your work. Last, you may fail to affirm the take-home message of your work and end strongly.

4.2.6.4 A Lack of Experience in Handling the Illustration of a Paper

> *An illustration is a visual editorial – it's just as nuanced. Everything that goes into it is a call you make: every color, every line weight, every angle.*
>
> *—Charles M. Blow (1970-)*

Since you are a new researcher who may lack adequate paper writing experience and knowledge, your brain may not be sufficiently conscious of making the right choices and creating the best designs for your illustrations [15]. As for making the right choices, you may fail to make the right choices between tables and figures, which include drawings, graphs, diagrams, and photographs [15]. As to creating the best designs, you may fail to create precise, clear, fluid, and familiar illustrations, though illustration goals are similar to language goals [15].

4.2.7 A Lack of Experience in Publishing a Paper

It's a dream until you write it down; then it's a goal.

—Emmitt Smith (1969-)

To succeed in your publishing endeavor as an emerging scholar, you need substantial experience in publishing a paper. Nevertheless, as a new Ph.D. student, researcher, and writer, you may lack adequate experience in publishing a paper, which is a Ph.D. pitfall that translates to multiple irreducible chicken-and-egg problems. In this vein, some of the publication-related Ph.D. pitfalls emanating from a lack of experience in publishing a paper are—but not limited to—itemized below.

! Attention: Publication-Related Ph.D. Pitfalls Emanating from a Lack of Experience in Publishing a Paper

- *A lack of experience in deciding on a conference and a journal.*
- *A lack of experience in how to satisfy the community of reviewers.*
- *A lack of experience in a paper rejection.*
- *A lack of experience in a paper acceptance.*

Next, each of the itemized Ph.D. pitfalls are discussed.

4.2.7.1 A Lack of Experience in Deciding on a Conference and a Journal

Publishing is a business. Writing may be art, but publishing, when all is said and done, comes down to dollars.

—Nicholas Sparks (1965-)

When deciding on a conference, you should consider its prestige and cost. In deciding on a journal, you better consider factors, such as quality, content, readership, length restrictions, publication lag, submission cost, authorship restrictions, editorial failings, open versus closed access, and journal citation reports [16, Ch. 14]. Despite such valuable pieces of advice, you—as a Ph.D. student and new researcher—may not have hands-on experience in choosing a conference/journal that fits your way of executing and reporting your research. Consequently, the entitled Ph.D. pitfall comes down to several irreducible chicken-and-egg problems you may face during the initial years of your Ph.D. study.

Below, we highlight one of such problems regarding the advice of experts that you need to have a particular class of journals/conferences in mind when you write a conference/journal paper [16, Ch. 11].

Because you may not know the majority of your research field's conferences and journals, you can hardly have a class of conferences/journals in mind while writing your conference/journal paper, so you will not be able to target the requirements and readership of that class of conferences/journals. As a result, you may not carefully choose conferences/journals that suit your way of conducting your research.

4.2.7.2 A Lack of Experience in How to Satisfy the Community of Reviewers

You can make anything with writing.

—C. S. Lewis (1898–1963)

As entering a new community often has its own rules and norms, entering any research community as a published author has its *unwritten rules and norms* since its many members may tend to reciprocate—as anonymous reviewers—what they have gone through as something proper.[3] As a Ph.D. student and new researcher lacking considerable paper publishing experience, your brain may not be *sufficiently conscious* of those rules and norms since you lack both knowledge and pragmatic experience. Hence, you may find it hard to satisfy the community of reviewers during the initial years of your Ph.D. study.

Although the best way to learn about the community of reviewers' expectations is through a hands-on endeavor of publishing regularly in top-tier conferences and journals so that you would have a nearly comprehensive understanding of how to satisfy your anonymous reviewers, you may not be qualified and experienced enough to regularly publish in top-tier conferences and journals during the initial years of your Ph.D. study. Therefore, the entitled Ph.D. pitfall boils down to several irreducible chicken-and-egg problems.

4.2.7.3 A Lack of Experience in a Paper Rejection

I love my rejection slips. They show me I try.

—Sylvia Plath (1932–1963)

Because you may not have a paper rejection experience as a new Ph.D. student and researcher, you may take the reviews of expert anonymous reviewers personally,

[3] Since the research community members' brains are causal biological machines, the researchers may tend to think in terms of what they have gone through as something right, for everyone's brain often falls into its own *neural experience*.

especially as a neuroscientifically subjective creature.[4] Apart from your subjectivity, your brain may not be *sufficiently conscious* of the many factors that can lead to a rejection of a paper, since you would be a new researcher who may lack considerable experience in publishing a paper. As a result, the entitled Ph.D. pitfall comes down to several irreducible chicken-and-egg problems you may face during the initial years of your Ph.D. study.

As a Ph.D. student and ambitious new researcher, you may aspire to develop new techniques by generating new ideas. However, the more your ideas diverge from your research field's mainstream trends of thinking, the harder it becomes to get them accepted since you have to devote more effort to convince senior researchers [16, Ch. 11], who would likely be your anonymous reviewers. As a Ph.D. student and new writer, you may not clarify what is new in your journal paper. Even if you successfully convey anything new in your article, what is new to you may not necessarily be new to anonymous expert reviewers. As a Ph.D. student and new researcher who is super busy, you may not have enough time for polishing and proofreading. Nevertheless, a typo-laden paper can easily be rejected by any anonymous reviewer.

4.2.7.4 A Lack of Experience in a Paper Acceptance

> *I've found most authors have the wrong mental picture of the process. Instead of a sprint, publishing is more like a marathon. Slow, steady and consistent action will get you your audience and success.*
>
> —*W. Terry Whalin (1953-)*

As a new researcher who may lack significant experience in publishing a paper, your brain may not be *sufficiently conscious* of the various factors that can inspire a conference/journal paper's acceptance. If you had considerable scholarly experience in getting published, your brain would easily be sufficiently conscious of those factors. However, you do not have considerable intellectual experience to be published as a new Ph.D. student, researcher, and writer. Therefore, the entitled Ph.D. pitfall boils down to many irreducible chicken-and-egg problems you may face during the initial years of your Ph.D. study.

As a new Ph.D. student and writer, you may fail to write with your readers—i.e., anonymous reviewers—in mind. However, since those reviewers potentially represent many other readers who may think differently from you [16, Ch. 11], they need to be convinced by you, who shall write with them in mind. To successfully

[4] Expert anonymous reviewers' comments—regardless of their tone—are always opportunities to improve the quality of your research, provided that you are calm enough to receive them objectively.

write your article, on the other hand, you need to start broadly to deal with the wide-reaching questions you will address; become more specific to focus on what you did; and end broadly to affirm the broad implications of your work [16, Ch. 11]. Nevertheless, as a Ph.D. student and new researcher, you may lack subject-matter expertise to start and end broadly.

4.3 Writing and Publication-Related Pitfalls During the Middle Years of a Ph.D.

> *Composition is for the most part an effort of slow diligence and steady perseverance, to which the mind is dragged by necessity or resolution, and from which the attention is every moment starting to more delightful amusements.*
>
> *—Samuel Johnson (1709–1784)*

During the middle years of your Ph.D. study, you will also run into various writing and publication-related Ph.D. pitfalls emanating from the inherently more challenging research, writing, and publication phase of your Ph.D. study, which is intertwined with your research, writing, and publication endeavor during the initial years of your Ph.D. study.

According to Wayne C. Booth, Gregory G. Colomb, Joseph M. Williams, Joseph Bizup, and William T. FitzGerald, who are the authors of a book entitled *The Craft of Research* [14], writing your concluded research is *multidimensional thinking*: *thinking with your readers* and *thinking for your readers* [14, Sec. 1.4, p. 36]. In doing so, you likely face various writing and publication-related Ph.D. pitfalls, which can be troublesome during the middle years of your Ph.D. study, for your Ph.D. (co-)supervisor's high expectations and a lack of advanced experience in publishing. Some of these pitfalls are—but not limited to—listed below and discussed next.

! Attention: Writing and Publication-Related Pitfalls During the Middle Years of a Ph.D.

- *Handling considerable boss pressure for a publication.*
- *A lack of experience in telling a story.*
- *A lack of experience in structuring a paper.*
- *A lack of advanced experience in writing a paper.*
- *A lack of advanced experience in publishing a paper.*

Next, we discuss each of the itemized Ph.D. pitfalls.

4.3.1 Handling Considerable Boss Pressure for a Publication

Mistakes and pressure are inevitable; the secret to getting past them is to stay calm.

—Travis Bradberry (1973-)

Because you have to justify your Ph.D. (co-)supervisor's investment in valuable time and funding, you as a Ph.D. student would naturally get a boss pressure for a publication as much as your Ph.D. (co-)supervisor would get from a funding agency, as detailed in Sect. 4.2.1. Because your Ph.D. (co-)supervisor would invest for multiple years in your Ph.D. research, you must justify the investment by delivering several novel results. Congruent with the requirements of a Ph.D., you are also expected to deliver many original results—that would be accepted by your research field's top-tier conferences and journals—during the initial and middle years of your Ph.D. study. Therefore, you would likely get considerable boss pressure for a publication throughout the middle years of your Ph.D. study.

Unless you are a new researcher who needs pressure to become focused and productive, your Ph.D. (co-)supervisor's considerable pressure and high expectations can be a challenge, especially when you are a curiosity-driven researcher; however, it is also a massive opportunity for rising to the challenges of working under pressure. As every perception is relative to one's mind experience, on the other hand, what you may perceive as a significant boss pressure may not be significant for a senior researcher who has already handled more pressure, since your brain would think in terms of what it has gone through. Nevertheless, as perceived by your mind, considerable pressure can induce immense stress and unsettling feelings of uncertainty, which can be detrimental to your productivity.

Handling considerable boss pressure requires significant experience in working under pressure. Nonetheless, as a Ph.D. student, you may not have significant experience working under pressure. Thus, the entitled writing and publication-related Ph.D. pitfall boils down to several irreducible chicken-and-egg problems.

4.3.2 A Lack of Advanced Experience in Telling a Story

Storytelling is the most powerful way to put ideas into the world.

—Robert McKee (1941-)

After inexhaustively handling the nitty-gritty details of the research you would conduct to obtain insightful results worth publishing, your next challenge would be publishing your research, especially as a new author with either no or limited publication experience. To succeed in this endeavor, many authors agree that you must be a good storyteller who writes a good paper telling a good story [16, 17]. To write a good paper that tells a good story, you should be able to make your idea sticky while properly structuring your paper's story [17].

According to Joshua Schimel, who is the author of a book entitled *Writing Science: How to Write Papers That Get Cited and Proposals That Get Funded* [17], the following are six major factors that make an idea sticky: ***s**implicity, **u**nexpectedness, **c**oncreteness, **c**redibility, **e**motional (i.e., curiosity), and **s**tory (**SUCCES**)* [17, p. 17]. Apart from the SUCCES elements, what is also integral to effective storytelling and science writing is the structure of a story, as noted below.

! Attention: The Four Core Structures of a Story [17, p. 27–32]

1. **OCAR structure**: **O**pening, **C**hallenge, **A**ction, and **R**esolution.
2. **ABDCE structure**: **A**ction, **B**ackground, **D**evelopment, **C**limax, and **E**nding.
3. **LD structure**: **L**ead and **D**evelopment.
4. **LDR structure**: **L**ead, **D**evelopment, and **R**esolution.

Despite the mentioned crucial guidance on generic storytelling, every novel paper has a unique story, which has to be written and narrated differently; it would have unique circumstances, motivation, and related works; it would focus on a unique problem that would be addressed theoretically, empirically, or experimentally; it would have unique solutions, results, and discussions; and it would have specific conclusions and research outlook. If you have to succeed in handling all these aspects of writing a good paper that tells a good story, you need to have advanced experience in telling a story. On the contrary, you may not have such advanced experience as a Ph.D. student and relatively new writer. Hence, the entitled writing and publication-related Ph.D. pitfall comes down to various irreducible chicken-and-egg problems.

4.3.3 A Lack of Advanced Experience in Structuring a Paper

If a man can group his ideas, then he is a writer.

—Robert Louis Stevenson (1850–1894)

During your research, you often have to wrestle with the various challenges of generating new ideas; formulating a novel problem; providing a solution by solving a novel problem; and validating your obtained solution using computer experiments or simulations. After persevering through these challenges, you will likely get new results worth publishing in one of your research field's top-tier journals. While preparing to start the hard work of writing your paper, you may wind up realizing that you have too many things to write in a paper whose coherent organization and structure you have no idea of. This is a writing and publication-related Ph.D. pitfall you would encounter during the middle years of your Ph.D. study, when you are too busy delivering original results.

Structuring a paper is an intellectually demanding activity that requires ten rules on *the principles of writing, the components of a paper, and the writing process* [18]. Handing all these rules to organize and structure a paper coherently requires advanced experience in structuring and writing a paper. As a Ph.D. student and relatively new writer, however, you may not have gathered advanced experience in structuring and writing a paper since you should first publish several novel journal papers to garner advanced experience in structuring and writing a paper. Therefore, the entitled writing and publication-related Ph.D. pitfall boils down to many irreducible chicken-and-egg problems, which you may likely face throughout the middle years of your Ph.D. study.

4.3.4 A Lack of Advanced Experience in Writing a Paper

> *One learns from books and example only that certain things can be done. Actual learning requires that you do those things.*
>
> *—Frank Herbert (1920–1986)*

As a relatively new writer who may lack advanced experience in writing a paper, you may fail to be concise, struggle to be fluid, and have problems in handling your prose. Thus, you may fail to make your idea sticky so that your writing flaws end up jeopardizing your chance of publishing it. This is among the many consequences of the entitled writing and publication-related Ph.D. pitfall, which is inevitable during the middle years of your Ph.D. study, giving rise to some of the writing-related Ph.D. pitfalls—but not limited to—listed below.

! Attention: Writing-Related Ph.D. Pitfalls due to a Lack of Advanced Experience in Writing a Paper

- *A lack of experience in being concise.*
- *A lack of experience in being fluid.*
- *A lack of experience in handling your prose.*

In what follows, we discuss each of the itemized Ph.D. pitfalls.

4.3.4.1 A Lack of Experience in Being Concise

> *In all pointed sentences, some degree of accuracy must be sacrificed to conciseness.*
>
> *—Samuel Johnson (1709–1784)*

When you succeed in *being clear and forthright*, your writing will be effectively *concise* while tight [15]. However, as a Ph.D. student and relatively new writer who may lack advanced writing experience, you can fail in being concise for several reasons. First, you may fail to eliminate redundancies; second, you may fail to reduce sentences to their simplest forms; and third, you may fail to identify and eliminate needless paragraphs and sections [15].

To be effectively concise in one's writing takes considerable effort, even for an experienced writer. To be consciously and effortlessly concise, you must write and publish several journal papers. On the contrary, as a Ph.D. student, you may not have written and published many journal papers. By giving rise to such irreducible chicken-and-egg problems, the entitled writing-related Ph.D. pitfall can be detrimental to your chance of getting it published, though your Ph.D. (co-)supervisor can always help you to be concise.

4.3.4.2 A Lack of Experience in Being Fluid

Everything flows and nothing abides; everything gives way and nothing stays fixed.

—Heraclitus (ca. 540-ca. 480 BC)

A lubricant that makes one's writing inform is *fluid language* [15]. For this reason, scientific language can be fluid, though its stylistic variation can be limited by the constraints of scientific writing [15]. Nonetheless, as a Ph.D. student and relatively new writer who may lack advanced writing experience, you may fail to be fluid by failing to vary your sentence rhythms and eliminate discontinuities [15]. In failing to vary your sentence rhythms, you may fail to vary your sentence openers, sentence lengths, sentence structures, and paragraph lengths [15]. In eliminating discontinuities, you may fail to make transitions between ideas and eliminate the needlessly complex typography [15].

To be a vividly fluid writer, you need to not only recognize the techniques of being fluid but also write and publish many journal papers. Nevertheless, as a Ph.D. student in the middle of a Ph.D. study, you may not have published several journal papers. Hence, the entitled writing-related Ph.D. pitfall boils down to various irreducible chicken-and-egg problems throughout the middle years of your Ph.D. study.

4.3.4.3 A Lack of Experience in Handling Your Prose

Prose is architecture, not interior decoration, and the Baroque is over.

—Ernest Hemingway (1899–1961)

To be a great writer of style and grace, you may need to write in different prose elegantly. As advocated by the authors of the book in [3], the most common style of many authors of articles is the *practical style*, which constitutes the language of manuals, memos, research reports, and term papers [2]. The job of these writers is mainly to serve the reader's immediate need to solve a problem, make a decision, conduct an inquiry, design or repair a machine, and so forth [1]. In the practical style, what initiates the writing is the reader's need, as opposed to the writer's desire to articulate something, which would be served by the writer who delivers timely materials [1].

As advocated by the authors of the books in [1, 2], *classic style (classic prose)* is a window onto the world, wherein what the reader has not yet noticed can be seen by a writer, who would orient the reader's gaze—by taking the form of a conversation—so that one can see it for oneself [1, 2]. In classic style, presentation is the purpose of writing, whose motive is disinterested truth, which succeeds—as measured by clarity and simplicity—when it aligns language with the truth [1, 2]. Different from the language that reveals it, the truth can be known by the writer of classic prose, who needs to present it [1, 2]. Before putting it into words and sentences, the truth is known by such a writer who is not using the occasion of writing to illuminate what one thinks nor have to argue for the truth [1, 2].

Despite their differences, classic and practical styles place a high value on clarity and directness; they are precise and efficient, though for quite different reasons [1]. As you keep garnering advanced experience in writing, you will recognize and consciously adopt different prose in English writing, especially classic prose. However, the elegant handling of the mentioned styles requires substantial writing experience, which you may lack throughout your Ph.D. study. Hence, the entitled writing-related Ph.D. pitfall comes down to several irreducible chicken-and-egg problems, which you may face throughout your Ph.D. study.

4.3.5 A Lack of Advanced Experience in Publishing a Paper

Our task is to communicate experience and ideas to others.

—Niels Bohr (1885–1962)

What affirm your originality and independence are your publications in top-tier journals, which are also timeless advertisements of your talents that would be widely known [10, p. 53–54], paving the way to your success in a Ph.D. and beyond. As a Ph.D. student with no or limited publication experience, nonetheless, you would be new to the publishing world and the unwritten rules of the global research enterprise. A lack of advanced experience in publishing a paper is thus a major Ph.D. pitfall that can specifically impact the middle years of your Ph.D. study, rendering some of the publication-related Ph.D. pitfalls—but not limited to—listed below.

! Attention: Publication-Related Ph.D. Pitfalls due to a Lack of Advanced Experience in Publishing a Paper

- *Some journal paper rejections.*
- *A lack of experience in aiming for a top-notch journal.*

Getting it published is a complicated theme, which is the subject of books and articles in [4, 15, 17, 19–21]. However, we focus below on the itemized Ph.D. pitfalls.

4.3.5.1 Some Journal Paper Rejections

> *Take rejection as an eye-opener. Take it as something that pushes you and makes you do things better.*
>
> *—Amit Sadh (1983-)*

Any paper—even a published one—can be rejected regarding rigorous expert scrutiny of infinite depth and breadth. In this vein, journal paper rejections are inevitable in your Ph.D. journey, especially during the middle years of a Ph.D. study, since you are expected to submit to top-tier journals in your research field. From a probabilistic standpoint, the likely rejection of your Ph.D. study papers boils down to a lack of significant novelty in your scholarly works and advanced experience in publishing a paper.

To succeed in your publication endeavor, you need to have considerable scholarly experience to get not only accepted but also rejected, which is an opportunity to uplift the quality of your research and paper. However, as a Ph.D. student, you can hardly garner considerable experience in publishing numerous papers. Thus, the entitled Ph.D. pitfall can cause many irreducible chicken-and-egg problems throughout your Ph.D. study.

4.3.5.2 A Lack of Experience in Aiming for a Top-Notch Journal

> *The greater danger for most of us lies not in setting our aim too high and falling short; but in setting our aim too low, and achieving our mark.*
>
> *—Michelangelo (1475–1564)*

As a new researcher lacking considerable paper publishing experience, your brain *may not be sufficiently sensible* to what it takes to get published in a top-tier journal. Specifically, you may not know the level of novelty needed to get accepted; the personal commitment required day in and day out; the level of persistence and determination needed; and the magnitude of ambition and expectancy you would need to propel yourself with. These constitute the entitled publication-related Ph.D. pitfall, which can impact your success in publishing top-tier journal papers during the middle years of your Ph.D. study, when you are expected to produce novel journal papers.

To succeed in publishing top-tier journal papers, you need to have significant experience in targeting top-tier journal papers. Nonetheless, as a Ph.D. student, you can hardly gather considerable experience in aiming for top-notch journal papers.

Furthermore, a good journal paper has always a story to tell [16, 17]. As a Ph.D. student and relatively new writer, however, you may fail to develop the story from your paper's beginning to its end [16, Ch. 11]. Therefore, the entitled publication-related Ph.D. pitfall translates to many irreducible chicken-and-egg problems.

4.4 Concluding Remarks

> *The biggest risk is not taking any risk. In a world that is changing quickly, the only strategy that is guaranteed to fail is not taking risks.*
>
> —*Mark Zuckerberg (1984-)*

This chapter detailed the several writing and publication-related Ph.D. pitfalls that are likely during the initial years of your Ph.D. study, such as boss pressure for a publication; neutralizing your subjectivities; compressing a voluminous work into a short paper; failing to see and manage the different layers of editing; realizing how much evidence, reasoning, and acknowledgment are enough; a lack of experience in writing a paper; and a lack of experience in publishing a paper. It also detailed various writing and publication-related Ph.D. pitfalls that are likely during the middle years of your Ph.D. study, such as handling considerable boss pressure for a publication, a lack of advanced experience in telling a story, a lack of advanced experience in structuring a paper, a lack of advanced experience in writing a paper, and a lack of advanced experience in publishing a paper.

Although we tried to discuss the most important writing and publication-related Ph.D. pitfalls inevitable during the initial and middle years of your Ph.D. study, our limited discussion—due to a space constraint—may not have covered the *full-blown depth and breadth* of issues intertwined with the discussed themes. For detailed insights and realities, we encourage you to reflect this chapter's references and discuss them with your network of peers.

References

1. Tomas FN, Turner M (2011) Clear and Simple as the Truth: Writing Classic Prose, 2nd edn. Princeton University Press, Princeton, NJ
2. Pinker S (2014) The Sense of Style: The Thinking Person's Guide to Writing in the 21st Century. Penguin, New York, NY
3. Strunk, Jr. W, White EB (2000) The Elements of Style, 4th edn. Longman, Needham Heights, MA
4. Rabinowitz H, Vogel S (eds.) (2009) The Manual of Scientific Style: A Guide for Authors, Editors, and Researchers. Academic, Burlington, MA
5. The University of Chicago Press (2017) The Chicago Manual of Style. The University of Chicago Press, Chicago, IL
6. Sword H (2012) Stylish Academic Writing. Harvard University Press, Cambridge, MA

7. Cron L (2016) Story Genius: How to Use Brain Science to Go Beyond Outlining and Write a Riveting Novel. Ten Speed Press, New York, NY
8. Zinsser W (2006) On Writing Well: The Classic Guide to Writing Nonfiction, 7th edn. HarperCollins, New York, NY
9. Silvia PJ (2019) How to Write a Lot : A Practical Guide to Productive Academic Writing, 2nd edn. American Psychological Association, Washington DC, DC
10. Feibelman PJ (2011) A PhD Is Not Enough!: A Guide to Survival in Science, Revised edn. Basic Books, New York, NY
11. Cicurel R, Nicolelis M (2015) The Relativistic Brain: How It Works and Why It Cannot Be Simulated by a Turing Machine. Kios Press, Sao Paulo
12. Nummenmaa L, Hari R, Hietanen JK, Glerean E (2018) Maps of subjective feelings. Proc Natl Acad Sci USA 115(37):9198–9203
13. Phillips EM, Pugh DS (2005) How to Get a PhD: A Handbook for Students and Their Supervisors, 4th edn. Open University Press, Berkshire
14. Booth WC, Colomb GG, Williams JM, Bizup J, FitzGerald WT (2016) The Craft of Research, 4th edn. The University of Chicago Press, Chicago, IL
15. Alley M (1996) The Craft of Scientific Writing, 3rd edn. Springer, New York, NY
16. Sternberg RJ, Sternberg K (2016) The Psychologist's Companion: A Guide to Professional Success for Students, Teachers, and Researchers, 6th edn. Cambridge University Press, New York, NY
17. Schimel J (2012) Writing Science: How to Write Papers That Get Cited and Proposals That Get Funded. Oxford University Press, New York, NY
18. Mensh B, Kording K (2017) Ten simple rules for structuring paper. PLoS Comput Biol 13:1–9
19. Day A (2007) How to Get Research Published in Journals, 2nd edn. Gower, Hampshire
20. El-Omar EM (2014) How to publish a scientific manuscript in a high-impact journal. Adv Digest Med 1:105–109
21. Mack CA (2018) How to Write a Good Scientific Paper. SPIE, Bellingham, Washington

Chapter 5
Management-Related Pitfalls of a Ph.D.

A disciplined mind leads to happiness, and an undisciplined mind leads to suffering.

—Dalai Lama (1940-)

Abstract Earning your Ph.D. requires the effective management of yourself, your Ph.D., and your Ph.D. (co-)supervisor throughout your Ph.D. study. Because such multi-dimensional management can be challenging for a Ph.D. student and new researcher, various management-related Ph.D. pitfalls undoubtedly arise throughout your Ph.D. study. Accordingly, this chapter discusses the management-related Ph.D. pitfalls that are inevitable during the initial, middle, and final years of your Ph.D. study so that you would be inspired to prepare yourself in advance.

Keywords Managing yourself · Managing your Ph.D. research · Managing your Ph.D. (co-)supervisor · Management-related Ph.D. pitfalls

5.1 Introduction

Management is doing things right; leadership is doing the right things.

—Peter Drucker (1909–2005)

As agreed by many researchers, your success in your Ph.D. study boils down to the effective management of yourself (as a new researcher and Ph.D. student), your Ph.D., and your Ph.D. (co-)supervisor. While anyone may tell you that you need to manage yourself, your Ph.D., and your Ph.D. (co-)supervisor effectively to earn your Ph.D., this is often *much easier said than done*, especially concerning the twenty-first-century academia's many unsettling realities and uncertainties that

T. M. Getu, *Pathways for Inspired Scientists and Engineers*,
https://doi.org/10.1007/978-3-032-06784-5_5

would impact you—directly or indirectly—as a new researcher and Ph.D. student. This is indeed the case for any Ph.D. student because separately managing yourself, your Ph.D., and your Ph.D. (co-)supervisor are intertwined with each other to the extent that jointly and effectively managing all three can be considerably and unpredictably challenging throughout your Ph.D. study.

Managing yourself as a new researcher and Ph.D. student is hardly easy for many reasons. First, you may not know or completely understand yourself as a new researcher and Ph.D. student who has to commit oneself for 3–5 years. Second, you can fail to identify and commit to your responsibilities as a new researcher and Ph.D. student. Third, you may knowingly or unknowingly *mismanage your time, your body, and your mind* throughout your Ph.D. studies. Fourth, you may not have yet figured out exactly what motivates and demotivates you. Fifth, you may fail to be cognizant of the psychological aspects of being a Ph.D. student. Sixth, you may tend to be more worried about your future rather than focused on the *(precious) present time*, especially when things are not turning out as you wish. Last, you may not have a strong purpose in science or a principled stance on it.

For any new researcher and Ph.D. student such as yourself, managing a Ph.D. is also far from a straightforward endeavor for many reasons. It can be challenging since you have to qualify for the usually high standards and requirements of Ph.D. research. It can also be challenging since you may need to overcome the challenges of your field's state-of-the-art research to deliver original results. It can also be challenging since you may need to overcome some (nearly) fundamental challenges of your research field to pave the way to new theories, techniques/algorithms, and frameworks. It can also be challenging since you have to deliver several cutting-edge publications throughout your Ph.D. study while working with your Ph.D. (co-)supervisor as a hand in glove. It can also be challenging since you have to qualify for high literary standards in almost all of your prospective conference and journal papers. All in all, it can also be challenging since you must surely overcome the many inevitable research-related, writing-related, publication-related, management-related, persona-related, and motivation-related Ph.D. pitfalls.

For any new researcher and Ph.D. student such as yourself, managing your Ph.D. (co-)supervisor can be challenging for many reasons. It can be challenging because you have to implicitly manage one's high expectations in research, novelty, writing, and publication. It can also be challenging because you have to implicitly manage multiple deadlines of conferences/symposiums, special issues of journals, magazines, and peer reviews. It can also be challenging because you have to implicitly manage your Ph.D. (co-)supervisor's research interests in disciplinary, interdisciplinary, multidisciplinary, and transdisciplinary fields. It can also be challenging because you have to implicitly manage your Ph.D. (co-)supervisor's interest in a competitive grant proposal preparation and writing. It can also be challenging because you have to implicitly manage your Ph.D. (co-)supervisor's behavior, which can sometimes vary and depend on your research performance. It can also be challenging because you have to implicitly manage your busy Ph.D. (co-)supervisor's mood[1] that can sometimes swing with some probability.

Pertaining to the reasons outlined above, various management-related Ph.D. pitfalls are inevitable during the initial, middle, and final years of your Ph.D. study. To this end, this chapter devotes considerable attention to understanding how and why management-related Ph.D. pitfalls are inevitable for the following reasons: first, to inform you in advance of the likely management-related Ph.D. pitfalls you would face in your Ph.D. study; second, to prepare you in advance so that you will not get bogged down by various management-related Ph.D. pitfalls inevitable during your Ph.D. study; third, to motivate you to turn proactively to the references cited by Chap. 5 and this book; fourth, to encourage you to discuss the management-related Ph.D. pitfalls with your network of peers; and last, to inspire you to develop *critical thinking, creative thinking, strategic thinking, probabilistic thinking, causal thinking, and independent thinking* capabilities, which you will need throughout your Ph.D. study.

For the reasons mentioned above, this chapter documents management-related Ph.D. pitfalls that are unavoidable during the initial, middle, and final years of your Ph.D. study. Concerning the initial years of your Ph.D. study, the management-related Ph.D. pitfalls we discuss in detail are failing to identify your responsibilities, struggling to manage your coursework and research work concurrently, failing to identify the responsibilities of your Ph.D. (co-)supervisor, failing to manage your Ph.D. (co-)supervisor, and failing to make your Ph.D. (co-)supervisor understand you. As to the middle years of your Ph.D. study, the management-related Ph.D. pitfalls we detail are failing to be mindful of the psychological aspects of the Ph.D. process, failing to work effectively under destruction, mismanagement of your time, mismanagement of your body and mind, and failing to manage the inevitable publication-related setbacks. As for the final years of your Ph.D. study, the management-related Ph.D. pitfalls we expose are having significant worries about your future, feelings of accomplishment and rest, a lack of funding, failing to get a big picture in the explanation of your Ph.D. dissertation, and a lack of awareness on the Ph.D. dissertation assessment criteria.

The remainder of this chapter is organized as follows. Section 5.2 exposes the management-related Ph.D. pitfalls inevitable during the initial years of your Ph.D. study. Section 5.3 details the management-related Ph.D. pitfalls unavoidable during the middle years of your Ph.D. study. Section 5.4 follows with the management-related Ph.D. pitfalls that are likely during the final years of your Ph.D. study. Eventually, Sect. 5.5 provides concluding remarks.

---

[1] While human behavior and mood can be unpredictable, especially in high-stress/high-demand environments like the twenty-first-century academia, there is a golden rule to manage any supervisor: *Deliver what is needed frequently and proactively.*

5.2 Management-Related Pitfalls During the Initial Years of a Ph.D.

It is not enough to be busy. So are the ants. The question is: What are we busy about?

—Henry David Thoreau (1817–1862)

Since effectively managing yourself, your Ph.D., and your Ph.D. (co-)supervisor can hardly be straightforward for a new researcher, you will likely encounter several management-related Ph.D. pitfalls throughout the initial years of your Ph.D. study. Some of these pitfalls are—but not limited to—listed below.

! Attention: Management-Related Pitfalls During the Initial Years of a Ph.D.

- *Failing to identify your responsibilities.*
- *Struggling to manage your coursework and research work concurrently.*
- *Failing to identify the responsibilities of your Ph.D. (co-)supervisor.*
- *Failing to manage your Ph.D. (co-)supervisor.*
- *Failing to make your Ph.D. (co-)supervisor understand you.*

In what follows, we discuss each of the itemized Ph.D. pitfalls.

5.2.1 *Failing to Identify Your Responsibilities*

If you take responsibility for yourself you will develop a hunger to accomplish your dreams.

—Les Brown (1945-)

During the initial years of your Ph.D. study, you would be new to the Ph.D. system and research enterprise, which may have its own unwritten rules and practices that your Ph.D. (co-) supervisor could enforce. This situation would make you swim in an ocean of new realities, since you would also be required to manage coursework and research work during the initial years of your Ph.D. study. Consequently, you may fail to identify your responsibilities to your Ph.D. (co-) supervisor, your Ph.D. program, your Ph.D. institution, your research community, and your research field.

Amid your swimming in an ocean of new realities and activities, you may lose contact with your Ph.D. (co-)supervisor while being overwhelmed by one's many demands. Unfortunately, this is *one of the many ways to fail in a Ph.D.* [1, p. 33–45] because your intellectual fitness as a Ph.D. student would be continually evaluated by your Ph.D. (co-)supervisor and your Ph.D. committee, to whom you would present your Ph.D. proposal. You may not also get a Ph.D. by failing to identify your scholarly responsibilities and act upon them for several reasons. First, because your scholarly responsibilities also shift with the nature and type of research problem you would be addressing, you may fail to identify your exact responsibilities, especially as a Ph.D. student in one's initial doctoral study years. Second, you may not precisely identify your scholarly responsibilities until you formulate an excellent research problem to solve. Third, scholarly endeavors are generally uncertain, even to experienced researchers of considerable research achievements. Accordingly, you may face the entitled management-related Ph.D. pitfall during the initial years of your Ph.D. study.

Once you fail to identify your responsibilities as a Ph.D. student, you will likely fail to execute them during the initial years of your Ph.D. study. If you fail to execute your responsibilities as a Ph.D. student, this is why a Ph.D. (co-) supervisor may likely push you. If you fail to execute your responsibilities as a Ph.D. student, this is precisely how you will fail—with a high probability—in becoming a Ph.D. candidate. On the other hand, if you have to grasp the depth and breadth of your responsibilities—to your Ph.D. (co-)supervisor, Ph.D. program, Ph.D. institution, research community, and field—as a Ph.D. student, you may need to garner considerable research experience. However, as a Ph.D. student, you would be only a new researcher with limited research experience, attesting to the several chicken-and-egg problems that the entitled management-related Ph.D. pitfall comes down to.

5.2.2 Struggling to Manage Coursework and Research Work Concurrently

I could only achieve success in my life through self-discipline, and I applied it until my wish and my will became one.

—Nikola Tesla (1856–1943)

During the initial years of your Ph.D. study, you are often expected to manage coursework and research work simultaneously while your intellectual fitness for your Ph.D. program is being assessed by your Ph.D. (co-)supervisor. On the contrary, because you have little or no experience in simultaneously managing coursework and research work during your bachelor's or master's study, you may end up having gigantic frustration that can often make you draw negative energy. Consequently, you may struggle to manage coursework and research work concurrently throughout the initial years of your doctoral study.

Attending several graduate courses and passing them satisfactorily while conducting novel research that can lead to a publication can be challenging during the initial years of your Ph.D. study, especially in the context of a typical North American Ph.D. system. Thus, you may struggle to manage coursework and research work for various reasons. First, you may lack advanced skills and experience in time management. Second, you may also lack advanced skills and experience in *time creation*, such as *hyper-focus*, *high effectiveness*, *imagination*, *collaboration* [2], *deep work* [3], and *depth of thinking* [4]. Third, you may also lack pragmatic management skills that can guide you through managing two things concurrently. Fourth, you may lack focus while managing two things simultaneously since multitasking can sometimes lead to a lack of productivity. Last, the limits of your intrinsic motivation and patience can be tested when you are subjected to the laborious work of managing coursework and research work concurrently daily.

During the challenging phase of the initial years of your Ph.D. study, you can reasonably be demotivated, frustrated, unenthusiastic, exhausted, isolated, and, at times, unhappy, which can all be the case for any Ph.D. student. No wonder you may struggle to manage coursework and research work simultaneously since this is mainly how to be a Ph.D. student and new researcher, from which you will grow into a senior researcher with scholarly experience and accomplishments. No wonder you may face the entitled Ph.D. pitfall since this is the most common management-related Ph.D. pitfall that was faced by several Ph.D. students of our time. Meanwhile, the effective management of dual scholarly responsibilities—such as coursework and research work—may require considerable management experience of multiple scholarly responsibilities. However, as a Ph.D. student and new researcher, you may lack significant management experience. Therefore, the entitled management-related Ph.D. pitfall comes down to various chicken-and-egg problems.

5.2.3 Failing to Identify the Responsibilities of Your Ph.D. (Co-)Supervisor

You cannot escape the responsibility of tomorrow by evading it today.

—Abraham Lincoln (1809–1865)

Your Ph.D. (co-)supervisor often has gigantic professional responsibilities in research, teaching, departmental/university services, community services and outreach, and professional services (to professional societies and one's profession). On the other hand, as a Ph.D. student, you are expected to independently deliver original research results published in top-tier journals that will constitute your Ph.D. dissertation while managing yourself, your Ph.D., and your Ph.D. (co-)supervisor. Hence, your world of Ph.D. studentship could not be more different from your Ph.D. (co-)supervisor's world of professorship and academia. However, while stressed by your possible lack of progress on your research, you may engage in wishful thinking that your Ph.D. (co-)supervisor should be hugely involved in your Ph.D. research, which is one of the effects of the entitled management-related Ph.D. pitfall.

Mindful of one's responsibilities, your prospective Ph.D. (co-)supervisor will likely become available when needed; read your work carefully and objectively; tend to be constructively critical, friendly, open, and supportive; lends expert knowledge and guidance in your research field; and have sufficient interest in your work to guide you ahead [1]. However, your Ph.D. (co-)supervisor may not have the willingness nor time to conduct a comprehensive literature survey for you, solve your challenging research problem, think through the depth and breadth of your research problem, formulate novel research problems for you, develop theorems and mathematical proofs for you, write software codes and run any computer simulations for you, write a draft journal/conference paper for you, and revise and submit a journal paper for you. Furthermore, due to the complexities and dynamics of a supervisor-student relationship, you may likely fail to understand the responsibilities of your Ph.D. (co-)supervisor, especially during the initial years of your Ph.D. study.

When you fail to identify the responsibilities of your Ph.D. (co-)supervisor as a manifestation of the entitled management-related Ph.D. pitfall, you will likely fail to engage one properly and manage one wisely, evoking incompetence and a lack of success in your Ph.D. study. On the other hand, if you are well aware of the responsibilities of your Ph.D. (co-)supervisor, it would help you legitimately expect things without compromising your independence as for the original research you must conduct and demonstrate during your Ph.D. study independently and convey *politely and respectfully* what you would expect out of your Ph.D. (co-)supervisor's professional supervision while discussing the progress of your research or drafted manuscripts. To completely understand your Ph.D. (co-)supervisor's responsibilities, you may need to effectively play the role of a professor supervising several Ph.D. students. On the contrary, you can hardly play the role of a professor while being a Ph.D. student and new researcher. Therefore, the entitled management-related Ph.D. pitfall boils down to several chicken-and-egg problems, which can bog down the initial years of your Ph.D. study.

5.2.4 Failing to Manage Your Ph.D. (Co-)Supervisor

A bad boss is like a disease of the soul.

—Chetan Bhagat (1974-)

As a neuroscientifically subjective creature, exactly like any human being, your Ph.D. (co-)supervisor can have one's likings, dislikes, preferences, interests, subjective attachments, behavior, mood, anger, frustration, etc. As a professor and competitive senior researcher, your Ph.D. (co-)supervisor can be super busy, under pressure, deadline driven, unhappy, grant oriented, paper oriented, plan oriented, high expectant, etc. Therefore, managing your Ph.D. (co-)supervisor is a complicated theme that you can likely fail at, especially during the initial years of your

Ph.D. study, since you may be super busy and absorbed while working under pressure.

As a Ph.D. student, you can fail to manage your Ph.D. (co-)supervisor since you have to manage one's high expectations in research and publication implicitly; challenging expert comments and criticisms that you may sometimes find stressful and unsettling; multiple deadlines for conferences/symposiums, special issues of journals, magazines, and reviews; quality of research as demonstrated by publications in top-tier journals, citation counts, impact on a research field, and depth and breadth of contributions; publication standard as demonstrated by top-notch conference venues, journal papers, magazines, and journals' special issues; award prestige of national and international awards; research interests in disciplinary, interdisciplinary, multidisciplinary, and transdisciplinary research fields; interest in a novel grant proposal writing; behavior that can vary anytime and dependent on your research performance; and personal mood that can swing whenever with some probability. Accordingly, during the initial years of your Ph.D. study, you may fail to manage your Ph.D. (co-)supervisor to the extent that you are nervous, stressed, and frustrated while feeling uncertain most of the time.

Despite the inevitability of the entitled management-related Ph.D. pitfall, you should be able to manage your Ph.D. (co-)supervisor all the time, regardless of any difficult situation you might be in, because your success in your Ph.D. largely depends on the management of not only your Ph.D. and yourself but also your Ph.D. (co-)supervisor. One of the best ways to manage any Ph.D. (co-)supervisor is to garner considerable research experience—by working and collaborating with many senior researchers and professors—that would guide your adaptive management and management wisdom. However, as a Ph.D. student, you can hardly gather significant research experience since you would be just a new researcher with limited knowledge and expertise. Therefore, the entitled management-related Ph.D. pitfall boils down to various chicken-and-egg problems, which you may face throughout the initial years of your Ph.D. study.

5.2.5 Failing to Make Your Ph.D. (Co-)Supervisor Understand You

Liberty means responsibility. That is why most men dread it.

—George Bernard Shaw (1856–1950)

Since your often-busy Ph.D. (co-)supervisor hardly has extra time to invest in understanding you, it is your unwritten responsibility to make your supervisor understand you completely. To succeed at this, you must first understand yourself very well. Unfortunately, while shouldering your responsibilities as a Ph.D. student and new researcher, you may sometimes go through a significantly emotional, daunting, and stressful experience. Consequently, your attention would often drift to your stressful situation to the extent that it may be hard to understand yourself completely. Should this situation recur frequently and you keep failing to understand yourself very well, your mind would often perceive and think relative to its inherent world—since *the human brain is relativistic* [5]—wherein you and yourself are not understood at all, making what would come out of your mind counterproductive to your progress during the initial years of your Ph.D. study. Accordingly, you would likely fail to understand yourself very well throughout the initial years of your Ph.D. study, making the entitled management-related Ph.D. pitfall inevitable throughout the initial years of your Ph.D. study.

There could be many reasons why you, as a Ph.D. student, may not entirely understand yourself as someone embarking on a scientific journey. First, as a new researcher, you may not know what you would be motivated and demotivated by. Second, you may not know how to spark and sustain your intrinsic motivation. Third, you may not know how to have too much fun out of your scientific research, scientific wonder, and unbounded curiosity. Fourth, you may not know if abundant research freedom would make you productive. Fifth, you may not know if you would be productive under intense supervisory pressure. Sixth, you may not know how to be productive under the pressure of your Ph.D (co-)supervisor. Seventh, you may not know if intense supervisory pressure would make you productive or depressed. Eighth, you may not know if plentiful research freedom would make you productive or lazy. Ninth, you may not know if you are curiosity or competition driven. Last, you may not know whether intense competition would make you anxious or work harder. As a result, you may fail to understand yourself thoroughly and fail to make your Ph.D. (co-)supervisor understand you.

When you fail to make your Ph.D. (co-)supervisor understand you, you can likely be an unfavorable Ph.D. student to work with, negatively impacting your likelihood of success in your Ph.D. study. During the initial years of your Ph.D. study, you may thus exert considerable effort to independently advance your research while engaging your Ph.D. (co-)supervisor as much as possible, putting effort to make one understand you. In the hope of making your Ph.D. (co-)supervisor understand you, meanwhile, one of the best ways to understand yourself entirely is to evaluate your progress, development, and transformation as a researcher—while becoming consistently cognizant of yourself and your surroundings—as you keep garnering significant research and intellectual experience. However, as a Ph.D. student and new researcher, you can hardly have significant research and intellectual experience. Therefore, the entitled management-related Ph.D. pitfall boils down to several chicken-and-egg problems, which you may come across throughout the initial years of your Ph.D. study.

We now proceed with the management-related pitfalls inevitable during the middle years of your Ph.D. study.

5.3 Management-Related Pitfalls During the Middle Years of a Ph.D.

Don't spend time beating on a wall, hoping to transform it into a door.

—Coco Chanel (1883–1971)

During the middle years of your Ph.D. study, you may already become a Ph.D. candidate who has already gathered practical early-stage research experience. Like the initial years of your Ph.D. study, you would also come across diverse management-related Ph.D. pitfalls due to your inherently more challenging research phase of the middle Ph.D. years and research engagement during the initial Ph.D. years. Some of these pitfalls are—but not limited to—itemized below.

! Attention: Management-Related Pitfalls During the Middle Years of a Ph.D.

- *Failing to be mindful of the psychological aspects of the Ph.D. process.*
- *Failing to work effectively under destruction.*
- *Mismanagement of your time.*
- *Mismanagement of your body and mind.*
- *Failing to manage publication-related setbacks.*

Next, we discuss each of the listed management-related Ph.D. pitfalls.

5.3.1 Failing to Be Mindful of the Psychological Aspects of the Ph.D. Process

Do not dwell in the past, do not dream of the future, concentrate the mind on the present moment.

—Buddha

During the initial years of your Ph.D. study, your enthusiasm for your Ph.D. and Ph.D. research can likely be high, like many Ph.D. students [1]. However, as time flies and your Ph.D. research gets naturally more challenging, your enthusiasm can dwindle quickly since you have to work on one major problem for multiple years, which may wind up feeling endless [1]. Once your enthusiasm diminishes significantly, you may lose the motivation that has brought you to a Ph.D. school to the extent that you would lose your primary purpose in a Ph.D. study. This is one of the major effects of the entitled management-related Ph.D. pitfall, which you may unconsciously face throughout the middle years of your Ph.D. study.

During the middle years of your Ph.D. study, you may have the most common anxiety among many Ph.D. students that "others are getting in first," where any other researcher would publish something on the same research topic as yours [1]. As you become more independent, this anxiety may force you to increase interest in your Ph.D. research so that you become more involved in your Ph.D. research for the sake of intrinsic interest and self-autonomy [1]. As you become more autonomous, you would implicitly transfer your dependence from your Ph.D. (co-)supervisor to your Ph.D. research, so you can effectively minimize the need for external approval or feedback [1]. As you depend more and more on your Ph.D. research while working persistently on your main problem, you will be isolated intellectually and socially, diminishing your initial enthusiasm [1]. As you become less enthusiastic during the middle years of your Ph.D. study, you may feel confused, fed up, and completely stuck [1]. As you become increasingly confused, fed up, and stuck, you may become frustrated to concentrate on your original research problem as it becomes more familiar and new follow-up ideas are generated [1]. As you become regularly frustrated, you would get to the point that your Ph.D. research work—like any other work—needs to be planned, developed, and completed within the time allotted when you exactly think that there is a job to be finished [1]. After one thing leads to another in the heat of your challenging intellectual journey as a relatively new researcher in the middle of a Ph.D. study, you may likely end up being mindless of the psychological aspects of the Ph.D. process, affirming the inevitability of the entitled management-related Ph.D. pitfall.

Being completely cognizant of the highlighted psychological aspects of the Ph.D. process is paramount for managing yourself and your Ph.D. study. To be completely mindful of the psychological aspects of the Ph.D. process, you have to go through the Ph.D. process and earn a Ph.D. degree. However, as a Ph.D. student, you have not yet gone through the Ph.D. process or earned a Ph.D. degree. Therefore, the entitled management-related Ph.D. pitfall boils down to various irreducible chicken-and-egg problems, which you may face throughout not only just the middle years of your Ph.D. study but also your entire Ph.D.

5.3.2 Failing to Work Effectively Under Destruction

> *Invention, it must be humbly admitted, does not consist in creating out of void, but out of chaos.*
>
> *—Mary Shelley (1797–1851)*

During the middle years of your Ph.D. study, you may likely be in a situation wherein you are required to effectively work under not only strict deadlines but also destruction. The inevitable destruction can emanate from intensive social engagement, wrong research results, unavoidable paper rejection, distracting literature, volunteering for the professional associations of your field, teaching responsibilities, administrative responsibilities (e.g., laboratory management), graduate funding applications, conference arrangements and travel, social media and societal interaction, and alike. The mental energy and attention you would give to such destruction would force you to snatch mental energy and attention from your Ph.D. research. As the destruction gets larger, so does the mental energy and attention you would snatch from your Ph.D. research. Accordingly, you would fail to work effectively under destruction, which underscores the inevitability of the entitled management-related Ph.D. pitfall that can occur throughout the middle years of your Ph.D. study.

Given the inevitability of the entitled management-related Ph.D. pitfall, you may be unable to work effectively under destruction for various reasons. First, you may not recognize destruction outright since it has many forms you may easily spend much of your time on without conscious awareness, dwarfing your productivity by getting you frustratingly nowhere. Second, you may not necessarily gravitate your life to the goals of your Ph.D. study, so your schedule can hardly be destruction-proof, and you can be easily distracted by temporary setbacks such as a paper rejection. Third, you may fail to choose and co-exist in the "right environment," such as the right environment for enriching personal and professional interactions; the right volunteering environment for your professional and personal development; the right conferences and journals for your Ph.D. research; the right scholarly research landscape of state-of-the-art disciplinary, interdisciplinary, multidisciplinary, and transdisciplinary research; the right professional network and collaboration, which can not only inspire but also challenge you; and so on. As a result, the highlighted destruction scenarios would make you fail—with a high likelihood—to work effectively under destruction, affirming the inevitability of the entitled management-related Ph.D. pitfall throughout the middle years of your Ph.D. study.

Because any destruction can easily draw your attention, energy, and time away from your Ph.D. research, it becomes tough to stay focused while getting distracted, even for a seasoned researcher, let alone for a relatively new researcher. Hence, you need to surmount the entitled management-related Ph.D. pitfall to succeed at working effectively under destruction. To effectively work under destruction, you may need significant research experience in managing destruction. Nonetheless, as a Ph.D. student and relatively new researcher, you can hardly garner significant research experience that would enable you to manage any destruction. Therefore, the entitled management-related Ph.D. pitfall boils down to several irreducible chicken-and-egg problems, which you may face throughout the middle years of your Ph.D. study.[2]

[2] You may also face this management-related Ph.D. pitfall throughout your entire Ph.D. study.

5.3.3 Mismanagement of Your Time

Lost time is never found again.

—Benjamin Franklin (1706–1790)

After transitioning from being a graduate student to a research student during the initial years of your Ph.D. study, many demands can further constrain your constrained time. These demands can be not only tasks and actions necessary for your Ph.D. research and professional growth but also irrelevant activities such as unnecessary socialization, phone calls, e-mails, appointments, and so on. Since these activities can have their urgency and feel-good factor, you may unknowingly or thoughtlessly engage with them at the expense of the necessary tasks and actions for your Ph.D. and professional growth. Consequently, you would end up mismanaging your precious time throughout the middle years of your Ph.D. study, affirming the inevitability of the entitled Ph.D. pitfall.

Concerning the inevitability of the entitled management-related Ph.D. pitfall, you may mismanage your valuable time for several reasons. First, you may unconsciously or thoughtlessly engage with unimportant and urgent tasks and actions such as calls from several national and international friends; frequent community engagement and volunteering; dealing with the urgent requests of others regularly; spending huge time on e-mails and frequent appointments; attending birthday parties regularly; invitations from several friends and acquaintances; and so forth. Second, you may unconsciously or thoughtlessly engage with unimportant and not urgent tasks and actions such as hanging out with several national and international friends; complaining and idling; (mindless) social media and its time-wasting *clickbaits*; reading and following gossip sites; overthinking and over planning; frequent Internet surfing, cable TV, Netflix, podcast, chatting, and gaming; intensive follow-up and reading of daily news; surfing several non-scientific forums; many excuses for coffee breaks, and social interactions; and so on. Third, you may unconsciously or thoughtlessly engage with various insignificant tasks and actions, even if they could be either urgent and important or not urgent and important. Last, the way you are *neuroscientifically wired*—according to your national culture and tradition—can likely force you to engage in unimportant, urgent, and insignificant socialization, which may feel right to your mind, though it can be a big time waster for your Ph.D. research, professional growth, and post-Ph.D. career.

While suffering from the entitled management-related Ph.D. pitfall, you may suffer from increased stress, loss of motivation, reduced morale, and missing opportunities unless you develop effective time management skills. One of the best ways to effectively manage your time is to gather significant research experience while conducting and managing a large body of research in time-constrained circumstances. Nevertheless, as a Ph.D. student and a relatively new researcher, you may not have succeeded in gathering significant research experience in a short period. Accordingly, overcoming the entitled management-related Ph.D. pitfall and effectively managing your precious time come down to several chicken-and-egg problems, which you may come across throughout the middle years of your Ph.D. study.[3]

5.3.4 Mismanagement of Your Body and Mind

If you want to conquer the anxiety of life, live in the moment, live in the breath.

—Amit Ray (1960-)

While breathlessly shouldering your gigantic Ph.D. responsibilities, such as delivering top-tier journal papers, as expected by your Ph.D. (co-)supervisor during the middle years of your Ph.D. study, you can be deeply absorbed to the extent that you would focus only on dealing with urgent tasks and actions, which can be important or unimportant regarding the objectives of your Ph.D. study. In being absorbed by the urgent tasks and actions that come your way, you would likely overlook and fail to regularly accommodate non-urgent and important tasks and actions, such as the effective management of your body and mind on a regular basis, which is indispensable for your health, happiness, and productivity. As a result, you would likely suffer from the detrimental effects of the entitled management-related Ph.D. pitfall throughout the middle years of your Ph.D. study.

Your productive time depends on the great state of your body and mind; otherwise, your time can hardly be productive irrespective of your laborious weekly hours and days. In sharp contrast, you may mismanage your body and mind for various reasons. First, whenever you are busy or breathless, your mind can be in an emergency mode so that it would simply want to deal with anything urgent coming your way, even if it could be unimportant and counterproductive for your Ph.D. research and professional career. Second, a lack of previous habits in managing body and mind may also make your mind believe that you would be academically successful if you keep working harder.[4] Third, your mind – in an attempt to reinforce its inherent belief system with mere hard work—may also think that you cannot afford to spend your already constrained valuable time on a regular gym (for your body) and meditation (for your mind) since you should be competitive with other excessively hardworking students of a Ph.D. system and a research enterprise. Fourth, as the main captain of yourself as a Ph.D. student and new researcher, your mind can be woefully inexperienced in successfully and jointly managing your body, mind, and Ph.D. research. Last, your mind can be negatively influenced by

[3] You may also face this management-related Ph.D. pitfall throughout your entire Ph.D. study.

[4] Contrary to the age-old mantra that hard work is the main success-driving factor, a couple of hours you regularly spend managing your body and mind can result in much more productive hours than the ones you would gain by simply working harder without caring for your body and mind.

the habits of successful researchers—in your surrounding or professional network—who may not spend proper time managing their bodies and minds.

Research is always a fun, entertaining, and exciting intellectual activity, provided that you can succeed in regularly managing your body and mind so that they can withstand the cognitive and physical demands of a persistent intellectual activity. Regularly managing your body and mind as a busy Ph.D. student and relatively new researcher requires a principled and objective mind, an excellent personal habit, effective time management, and extensive experience in jointly managing one's research, body, and mind. Nonetheless, as a Ph.D. student and a relatively new researcher, you may lack effective time management skills and extensive experience in jointly managing your research, body, and mind. Therefore, surmounting the entitled management-related Ph.D. pitfall and effectively managing your body and mind boil down to several chicken-and-egg problems, which you may come across throughout the middle years of your Ph.D. study.[5]

5.3.5 Failing to Manage Publication-Related Setbacks

Let go of your mind and then be mindful.

—Rumi (1207–1273)

As discussed in Chap. 4, any paper—even a published one—can be rejected concerning rigorous expert scrutiny of infinite depth and breadth.[6] During the middle years of your Ph.D. study, you may likely face some paper rejections, which can cause depression, immense demotivation, significant loss of research interest, and erosion of self-belief, especially as a relatively new researcher. While these feelings and psychological impacts are humanistic and common for a Ph.D. student, should they be felt for a few days to a week, you may fail to manage the inevitable publication-related setbacks for multiple weeks, especially throughout the middle years of your Ph.D. study.

For several reasons, you may fail to manage the inevitable publication-related setbacks, especially throughout the middle years of your Ph.D. study. First, as a Ph.D. student and a relatively new researcher, you may be new to a peer-review system and research enterprise. Second, you may subscribe to the myopic view that science is a quantifiable and discreet destination rather than a never-ending process and purpose. Third, considering your lack of research experience and the uncertainty

[5] You may also be confronted by this management-related Ph.D. pitfall—like many junior and senior researchers—throughout your entire Ph.D. study.

[6] It is worth underscoring that rejection is an unavoidable thing even in the life of a distinguished senior researcher, let alone a relatively new researcher.

of a scientific journey, you may not have principled long-term research goals and vision to spare you from immense disappointment from temporary setbacks such as a paper rejection. Fourth, the rejection you may face can severely hurt your self-confidence, especially when it happens after your persistence and hard fight for acceptance. Last, the human brain is fundamentally *reward centric*, making any rejection hard to swallow for our progress, even if we all know that nothing in this universe is perfect. Accordingly, the entitled management-related Ph.D. pitfall would be inevitable throughout the middle years of your Ph.D. study.

Your success in the middle years of your Ph.D. study hinges on your not only enthusiastic and curiosity-driven research but also your resilience in effectively managing the inevitable publication-related setbacks. Succeeding at this, meanwhile, requires considerable scholarly experience in research, publication, and leadership while overcoming any probable feelings of self-defeat, depression, loss of research interest, and diminishing inherent motivation. However, as a Ph.D. student and relatively new researcher, you may not have yet garnered significant scholarly experience in research, publication, and leadership. Therefore, surmounting the entitled management-related Ph.D. pitfall and effectively managing the inevitable publication-related setbacks boils down to several chicken-and-egg problems, which you may face throughout the middle years of your Ph.D. study.

We now move on to the management-related pitfalls evident during the final years of your Ph.D. study.

5.4 Management-Related Pitfalls During the Final Years of a Ph.D.

Don't watch the clock; do what it does. Keep going.

—Sam Levenson (1911–1980)

Concerning the realities of numerous Ph.D. pitfalls evident during the initial and middle years of your Ph.D. study, being a Ph.D. student may sometimes feel like being a passenger of a ship sailing against oceanic winds, where any passenger can likely feel total relief and joy when the ship eventually arrives at a region full of sunlight and blue skies. Analogously, when you make it to the final years of your Ph.D. study, you may feel accomplishment, rest, and confidence. However, in an often unpredictable and demanding system like a Ph.D., anything can go wrong anytime, including during the final years of your Ph.D. study, which shall be a year of hope, vigilance, hard work, thoughtfulness, mature engagement, and wise management that can finally pave the way to your successful Ph.D. defense and deposit of the final version of your prospective Ph.D. dissertation.

Aiming to inform and empower you toward your Ph.D. degree, this section presents the management-related Ph.D. pitfalls inevitable during the final years of your Ph.D. study. Some of these Ph.D. pitfalls are—but not limited to—itemized below.

! Attention: Management-Related Pitfalls During the Final Year of a Ph.D.

- *Having significant worries about your future.*
- *Feelings of accomplishment and rest.*
- *A lack of funding.*
- *Failing to get a big picture in the explanation of your Ph.D. dissertation.*
- *A lack of awareness of the Ph.D. dissertation assessment criteria.*

In what follows, each of the itemized Ph.D. pitfalls is discussed individually.

5.4.1 Having Significant Worries About Your Future

As for the future, your task is not to foresee it, but to enable it.

—Antoine de Saint Exupery (1900–1944)

As your Ph.D. approaches completion, possibly after a hard fight and some episodes of frustration, it is a natural phenomenon that your mind would look to your post-Ph.D. future throughout the final years of your Ph.D. study. Such a cognitive tendency, coupled with any unpredictably bad situation in contemporary academia, industry, and government, often makes anyone worry considerably about one's post-Ph.D. future. The possibility of a dire job market—in academia, industry, and government—can also make each of your colleagues worry significantly about their futures to the extent that everyone ends up attracting worries from each other. As a result, many things may appear justified, so you would be forced to harbor substantial worries about your future continuously. Accordingly, the entitled management-related Ph.D. pitfall will be inevitable throughout the final years of your Ph.D. study.

Having immense and recurrent worries about the future can easily destroy your focus and attention on not only your Ph.D. but also your final-year Ph.D. research, which is essential for the sound conclusion of your prospective Ph.D. dissertation. Even if you are well aware of this, you may have significant and recurrent worries about your future for various reasons. First, your mind can fundamentally be stressed and driven by uncertainty, which it may always aim to minimize. Second, you can be knowingly or unknowingly influenced by the engagement and attitude of your colleagues and Ph.D. students. Third, you may likely be convinced that thinking about your future before your Ph.D. graduation is a proactive strategy for taking care of your future, since your mind is also uncertainty driven. Fourth, you can also be influenced by your social media and professional network that you would implicitly be worried about your post-Ph.D. future. Last, your accomplishments in your Ph.D. research and the hyper-competitive academic ecosystem may regularly alarm your mind to worry considerably about your future. Consequently, the entitled management-related Ph.D. pitfall would be unavoidable throughout the final years of your Ph.D. study.

Because the benefits of shifting your mind and attention toward the future often come at the cost of losing your attention in the present moment, you should be well aware of this fundamental *cognitive tradeoff* before engaging with anything relevant to your future. To this end, you shall put *first things first*—like many highly effective people—and focus on completing your Ph.D. rather than simply getting lost while worrying about your future. To not suffer from significant worries about your future, you need to be a settled research scholar nearly ready to be an established researcher in your research field. As a final-year Ph.D. student and relatively new researcher, however, you may not be nearly ready to be an established researcher in your research field, except for some exceptional and anecdotal cases. Therefore, the entitled management-related Ph.D. pitfall boils down to several irreducible chicken-and-egg problems, which you may face throughout the final years of your Ph.D. study.

5.4.2 Feelings of Accomplishment and Rest

> *You cannot swim for new horizons until you have the courage to lose sight of the shore.*
>
> *—William Faulkner (1897–1962)*

As your Ph.D. nears its end after overcoming various pitfalls of the initial and middle years of your Ph.D. study, the pleasant feelings of accomplishment and rest may likely fill your head during the final years of your Ph.D. study. As you become more certain about accomplishing your Ph.D. dissertation, the feelings of rest and accomplishment will strengthen. Despite such a feeling getting pleasantly stronger and stronger, you have not yet submitted nor successfully defended your Ph.D. dissertation. Hence, you still have too much work to wrap up before being eligible for your Ph.D. defense, though your brain would somehow be deceived by its own inherent fundamental neural processing—incompletely understood fundamental phenomena—that is producing your recurring feeling of accomplishment and rest. However, this is precisely a management-related Ph.D. pitfall common throughout the final years of your Ph.D. study.

Even if you may be well aware that feelings of accomplishment and rest can negatively impact the completion of your Ph.D. study, you may continually harbor such a feeling throughout the final years of your Ph.D. study for several reasons. First, feelings of accomplishment and rest can be your mind's fundamental way of minimizing uncertainty and stress, which it can face during the initial and middle years of your Ph.D. study. Second, feelings of accomplishment and rest can create an unjustified positive feeling that can entice your mind and attention away from your progressing Ph.D. dissertation. Third, you may fail to repeatedly remind yourself that you have not yet submitted your Ph.D. dissertation, which should be rigorously defended in front of a panel of subject-matter experts. Fourth, trying to minimize any feeling of uncertainty and stress, your mind may overlook the fact that you still need to instill *burning desire and persistence* into your biological neural circuits so that you would be able to successfully arrive at the defense of your Ph.D. research and submission of your final Ph.D. dissertation. Last, you can be negatively influenced by the environment created by your professional network and colleagues of final-year Ph.D. students so that you may tend to socialize more and work much less while feeling accomplishment and rest.

When you feel that you are done and accomplished, you will tend to research much less to the degree that you would evoke considerable *vacant time* into your ginormous *biological neural circuits* embedded inside your head. As a result, the abundant vacant time would drag your entire mind into distraction and possibly irrelevant social engagements, which can hardly do justice to your final-year Ph.D. research. Hence, it is critically important to be well aware in advance of the detrimental effects of the entitled management-related Ph.D. pitfall so as not to put your unfinished Ph.D. dissertation at stake. Meanwhile, you would unlikely have feelings of accomplishment and rest, provided that you have a bunch of notable accomplishments and a burning desire to earn much more in science and innovation. As a first-year Ph.D. student and relatively new researcher, however, you may not have a bunch of notable accomplishments, though everything is relative, including the word notable. Therefore, the entitled management-related Ph.D. pitfall comes down to several irreducible chicken-and-egg problems, which you may face throughout the final years of your Ph.D. study.

5.4.3 A Lack of Funding

Money is a terrible master but an excellent servant.

—P.T. Barnum (1810–1891)

After going through and surmounting several Ph.D. pitfalls unavoidable during the initial and middle years of your Ph.D. study, you may tend to think that you are already there—because you would be in your final Ph.D. year—and hence, you should take everything easy. However, at any time during the final year of your Ph.D. study, your Ph.D. (co-)supervisor may be forced to break the cold news to you that your Ph.D. project will soon run out of funding. In such a likely circumstance, you would be forced to sustain yourself financially since you cannot halt your Ph.D. study at its eleventh hour while suffering from a lack of funding throughout the final year of your Ph.D. study. This is how you may come across the entitled management-related Ph.D. pitfall, since your Ph.D. scholarship would likely be a constrained grant.

During the final years of your Ph.D. study, you may be subjected to a lack of funding for several reasons. First, a Ph.D. scholarship is often a hardly sufficient grant that can enable you to accumulate abundant savings, especially in a world of consistently rising inflation, for rainy days. Second, many Ph.D. students in many universities often aim for a few available fellowships that are usually hyper-competitive across the entire college/school and university. Third, to release yourself from a possibly exhausting, uncertain, and stressful Ph.D. life, you may sometimes take it easy and go out to enjoy yourself very well, which can deplete your savings, if you succeed in saving. Last, because your Ph.D. (co-)supervisor may sponsor several researchers using the grant of one's big research project, one can run out of funding—with some probability—before completing your Ph.D. study. Accordingly, you may face the entitled management-related Ph.D. pitfall during the final years of your Ph.D. study.

Since a lack of funding would cause you substantial stress, demotivation, frustration, and destruction, you would tend to think more and more about your financial troubles, forcing you to focus less and less on your miscellaneous Ph.D. research. Consequently, you and your distracted mind would fall into a recursive trap of *unproductive times and unproductivity*, regardless of the hours you may stare at your computer. Such unproductivity can make your final-year Ph.D. life wickedly miserable because you need no more bad news—in an already long and arduous intellectual journey—that can hinder your ability and success in overcoming a lack of funding. To overcome a lack of funding,[7] meanwhile, significant experience in managing funding shall be garnered by you, who shall also have the capability and freedom to apply for your funding. Nevertheless, as a Ph.D. student and relatively new researcher, you may not have the capability and freedom to apply for your funding nor significant experience in managing funding. Therefore, you may face the entitled management-related Ph.D. pitfall during the final year of your Ph.D. study.

5.4.4 Failing to Get a Big Picture in the Explanation of Your Ph.D. Dissertation

> *If you just focus on the smallest details, you never get the big picture right.*
>
> *—Leroy Hood (1938-)*

While toiling to publish some journal papers that would pave the way to the final year of your Ph.D. study, you would certainly be immersed—both personally

[7] To overcome a lack of funding, you shall save money regularly and target various fellowships while making yourself a competitive Ph.D. student.

and neuroscientifically—in the nitty-gritty details of your Ph.D. research, which comprise the building blocks of the *scholarly house* you would be building through your Ph.D. research. Even though your mind would spend much of your time embedded in such building blocks, what is the most visible and appealing from the outside—regardless of the reasons—is the scholarly house to be built by you and, hence, the *big picture*. Due to the high expectations and publication-oriented goals of most Ph.D. systems, you could merely be immersed in various research details throughout the years to the extent that you would fail to get a big picture that would inspire the explanation of your Ph.D. dissertation. This is precisely the entitled management-related Ph.D. pitfall you may face throughout the final years of your Ph.D. study.

Because of the entitled management-related Ph.D. pitfall, you would likely struggle to get a coherent Ph.D. dissertation out of your research for many reasons. First, you may lack sufficient cognitive training from several viewpoints in top-down and bottom-up thinking and imagination. Second, you may lack considerable cognitive experience in top-down and bottom-up thinking and imagination from various vantage points. Third, you may lack depth and breadth of subject-matter expertise that would enable you to see the big picture, points of convergence/divergence, and the gap in the existing literature of your research field. Last, you may not have adequate time to ask the why, what, and how of everything you may read and investigate. Accordingly, the entitled management-related Ph.D. pitfall[8] would be inevitable throughout the final years of your Ph.D. study.

Since the experts who are members of your Ph.D. dissertation committee often start with the big picture and keep asking for additional clarifications/details, failing to get the big picture of your Ph.D. research would make you *psychologically weak* for your prospective Ph.D. defense. Meanwhile, getting a big picture out of many pertinent details—such as your doctoral research works over the years—requires substantial research experience, cognitive training/experience in bottom-up and top-down thinking, as well as imagination from various vantage points, which need significant expertise in your subject matter. As a Ph.D. student and relatively new researcher, however, you may lack significant expertise in your research field, considerable research experience, and significant cognitive training/experience in *bottom-up and top-down thinking and imagination*. Therefore, the entitled management-related Ph.D. pitfall boils down to various chicken-and-egg problems, which you may be confronted with throughout the final years of your doctoral study.

[8] Note that this entitled management-related Ph.D. pitfall can also be a research-related Ph.D. pitfall.

5.4.5 A Lack of Awareness of the Ph.D. Dissertation Assessment Criteria

Educational success should be measured by how strong your desire is to keep learning.

—Alfie Kohn (1957-)

Either intentionally or unintentionally, a Ph.D. dissertation assessment criteria can be vague in many schools and universities to the extent that you may not precisely know how the outlined assessment criteria—if any—can be applied to evaluate your unique Ph.D. research and problems. As a result, you may not exactly know how your prospective Ph.D. dissertation would be evaluated, so you would be in the dark throughout the preparation of your prospective Ph.D. dissertation while possibly ending up with several problems in presentation, clarity, unification, coherence, organization, discussion, data analysis, and alike. This is how the entitled management-related Ph.D. pitfall can impact the final years of your Ph.D. study.

You may not completely know your university and school's Ph.D. dissertation assessment criteria for various reasons. First, both your Ph.D. (co-)supervisor and your school's administrators can be too busy to exhaustively explain the school's Ph.D. dissertation assessment criteria, though supervisors may tend to state their scholarly expectations at the beginning. Second, even if you may be well aware of your school's assessment criteria for a Ph.D. dissertation, they can be subjective to the judgment and taste of an expert researcher who is a member of your Ph.D. committee. Third, even if you may feel that you comprehend the Ph.D. assessment criteria of your school and university, you may not know how they can be applied to your unique research problem and Ph.D. dissertation. Fourth, Ph.D. dissertation assessment criteria such as novelty, significance, and impact can always be relative—never absolute, making them dependent on an expert evaluator's taste, preference, and experience. Last, even if you may have read about your school and university's Ph.D. dissertation evaluation criteria, you may not know precisely the (unwritten) assessment criteria and their implementation unless you go through your Ph.D. defense. Consequently, the entitled management-related Ph.D. pitfall would be inevitable throughout the final years of your Ph.D. study.

To effectively surmount the impacts of the entitled management-related Ph.D. pitfall without getting bogged down by it, you are advised to put sound effort so that you would get a glimpse of the Ph.D. dissertation assessment criteria of your school/university as early as possible since the sooner is always the better. Meanwhile, the best way to be well aware of the Ph.D. dissertation assessment criteria is to serve on many Ph.D. committees and experience the overall process yourself. As a Ph.D. student and relatively new researcher, you haven't earned your Ph.D. yet to be on many Ph.D. committees. Hence, the entitled management-related Ph.D. pitfall boils down to several irreducible chicken-and-egg problems, which you may likely face during the final years of your Ph.D. study.

5.5 Concluding Remarks

The price of greatness is responsibility.

—Winston Churchill (1874–1965)

This chapter presented management-related Ph.D. pitfalls that are inevitable during the initial, middle, and final years of your Ph.D. study. Regarding the initial years of your Ph.D. study, the management-related Ph.D. pitfalls we detailed are failing to identify your responsibilities, struggling to manage your coursework and research work concurrently, failing to identify the responsibilities of your Ph.D. (co-)supervisor, failing to manage your Ph.D. (co-)supervisor, and failing to make your Ph.D. (co-)supervisor understand you. Regarding the middle years of your Ph.D. study, the management-related Ph.D. pitfalls we detailed are failing to be mindful of the psychological aspects of the Ph.D. process, failing to work effectively under destruction, mismanagement of your time, mismanagement of your body and mind, and failing to manage the inevitable publication-related setbacks. Regarding the final years of your Ph.D. study, the management-related Ph.D. pitfalls we detailed are having significant worries about your future, feelings of accomplishment and rest, a lack of funding, failing to get a big picture in the explanation of your Ph.D. dissertation, and a lack of awareness of the Ph.D. dissertation assessment criteria.

Although we attempted to discuss the most important management-related Ph.D. pitfalls inevitable during the initial, middle, and final years of your Ph.D. study, our limited discussion—due to a space constraint—may not have covered the *full-blown depth and breadth* of issues intertwined with the discussed themes. For detailed insights and realities, we encourage you to reflect this chapter's references and discuss them with your network of peers.

References

1. Phillips EM, Pugh DS (2005) How to Get a PhD: A Handbook for Students and Their Supervisors, 4th edn. Open University Press, Berkshire
2. Anandarajan A (ed) (2010) e-Research Collaboration: Theory, Techniques and Challenges. Springer, Berlin
3. Newport C (2016) Deep Work: Rules for Focused Success in a Distracted World. Grand Central Publishing, New York, NY
4. Paul RW, Elder L (2002) Critical Thinking: Tools for Taking Charge of Your Professional and Personal Life. Prentice Hall, Upper Saddle River, NJ
5. Cicurel R, Nicolelis M (2015) The Relativistic Brain: How It Works and Why It Cannot Be Simulated by a Turing Machine. Kios Press, Sao Paulo

Chapter 6
Persona and Motivation-Related Pitfalls of a Ph.D.

A hallmark of those who achieve greatness is the discovery that they can control the level of motivation they feel by better directing their own minds.

—Brendon Burchard (1977–)

Abstract The many management-related Ph.D. pitfalls, in tandem with the several research-related, writing-related, and publication-related Ph.D. pitfalls, cause numerous persona and motivation-related Ph.D. pitfalls that are inevitable during the initial and middle years of your Ph.D. study. Accordingly, this chapter details the persona and motivation-related Ph.D. pitfalls unavoidable during the initial and middle years of your Ph.D. study so that you would be inspired to prepare yourself in advance.

Keywords Persona-related Ph.D. pitfalls · Motivation-related Ph.D. pitfalls · Inevitability of persona and motivation-related Ph.D. pitfalls

6.1 Introduction

It always seems impossible until it's done.

—Nelson Mandela (1918–2013)

The high intellectual standards and many uncertainties of doctoral research lead to several research-related Ph.D. pitfalls, especially during the initial and middle years of your Ph.D. study. While working to overcome the many research-related Ph.D. pitfalls during the initial years of your Ph.D. study, you may take everything easy; you may suffer from imposter syndrome; you may avoid undertaking essential challenges; you may wish to succeed as soon as possible, with minimal endeavors; and you may fail to understand yourself very well as a new researcher and Ph.D. student. These persona-related Ph.D. pitfalls can reinforce your negativity and bad feelings that you may feel you are working long hours and achieving nothing; having

T. M. Getu, *Pathways for Inspired Scientists and Engineers*,
https://doi.org/10.1007/978-3-032-06784-5_6

no inherent motivation, positive energy, and drive; having no research momentum to build upon; and being overwhelmed by many papers and results. These are the inevitable motivation-related Ph.D. pitfalls that can be caused by not only research-related Ph.D. pitfalls but also writing-related, publication-related, and management-related Ph.D. pitfalls.

While working to surmount the several research-related Ph.D. pitfalls during the middle years of your Ph.D. study, you may follow the crowd for nothing; spend much time comparing yourself with others; get distracted by wrong results, though they are naturally a part of the research process; lack focus when things are not working as you wish; and you may draw enormous negative energy when things are not working as planned. Inevitable during the middle years of your Ph.D. study, these persona-related Ph.D. pitfalls bring forth motivation-related Ph.D. pitfalls that are inevitable during the middle years of your Ph.D. study, such as struggling to stay motivated, frustration with the Ph.D. process, staying motivated when rejected, working long hours alone, and feelings of exhaustion. It is worth noting that these inevitable motivation-related Ph.D. pitfalls can be caused by not only research-related Ph.D. pitfalls but also writing-related, publication-related, and management-related Ph.D. pitfalls.

Although the inevitable persona and motivation-related Ph.D. pitfalls can be caused by research-related, writing-related, publication-related, and management-related Ph.D. pitfalls, we may not be able to infer exactly how persona and motivation-related Ph.D. pitfalls arise. This is because they can be caused not only by one of the mentioned four Ph.D. pitfalls but also by a combination of them. In this vein, the inevitable research-related Ph.D. pitfalls coupled with your lack of writing and publication experience give rise to various writing and publication-related Ph.D. pitfalls, particularly during the initial and middle years of your Ph.D. study. These inevitable pitfalls, along with the unavoidable research-related Ph.D. pitfalls, bring forth several management-related Ph.D. pitfalls throughout your entire Ph.D. study. Alas, these Ph.D. pitfalls, in conjunction with the many research-related, writing-related, and publication-related Ph.D. pitfalls, cause numerous persona and motivation-related Ph.D. pitfalls that are inevitable throughout the initial and middle years of your Ph.D. study.

As to the inevitability of various persona and motivation-related Ph.D. pitfalls, this chapter offers immense attention to how and why they happen for the following reasons: first, to inform you in advance of the probable persona and motivation-related Ph.D. pitfalls you would face in your Ph.D. study; second, to prepare you in advance so that you will not get bogged down by the inevitability of various persona and motivation-related Ph.D. pitfalls during your Ph.D. study; third, to motivate you to turn proactively to the references cited by Chap. 6 and this book; fourth, to encourage you to discuss the persona and motivation-related Ph.D. pitfalls with your network of peers; and last, to inspire you to develop *critical thinking, creative thinking, strategic thinking, probabilistic thinking, causal thinking, and independent thinking* capabilities, which you will need throughout your Ph.D. study.

For the abovementioned reasons, this chapter presents detailed and critical arguments on the various persona and motivation-related Ph.D. pitfalls inevitable throughout the initial and middle years of your Ph.D. study. As for the initial years of your Ph.D. study, this chapter discusses persona-related Ph.D. pitfalls such as taking everything easy, failing to understand yourself very well, not being ready to embrace challenges, suffering from *imposter syndrome*, and early expectations of success. As for the initial years of your Ph.D. study, this chapter also discusses motivation-related Ph.D. pitfalls such as being overwhelmed, working long hours alone, a lack of momentum, a lack of motivation, and a lack of positive energy and drive. As to the middle years of your Ph.D. study, this chapter details persona-related Ph.D. pitfalls such as a lack of focus when things are not working, drawing negative energy when things are not working, getting distracted by wrong results, following the crowd, and comparing yourself with others. As to the middle years of your Ph.D. study, this chapter also details motivation-related Ph.D. pitfalls such as working long hours alone, staying focused and motivated when rejected, struggling to stay motivated, frustration with the Ph.D. process, and feelings of exhaustion.

The remainder of this chapter is organized as follows. Section 6.2 details the persona-related pitfalls inevitable during the initial years of your Ph.D. study. Section 6.3 follows with the persona-related pitfalls unavoidable during the middle years of your Ph.D. study. Section 6.4 documents the motivation-related pitfalls probable during the initial years of your Ph.D. study. Section 6.5 discusses the motivation-related pitfalls likely during the middle years of your Ph.D. study. Finally, Sect. 6.6 ends this chapter with concluding remarks.

6.2 Persona-Related Pitfalls During the Initial Years of a Ph.D.

Personality has power to uplift, power to depress, power to curse, and power to bless.

—*Paul P. Harris (1868–1947)*

This section discusses the persona-related Ph.D. pitfalls that are likely during the initial years of your Ph.D. study, regardless of your research field. Some of these Ph.D. pitfalls are—but not limited to—itemized below.

! Attention: Persona-Related Pitfalls During the Initial Years of a Ph.D.

- *Taking everything easy.*
- *Failing to understand yourself very well.*
- *Not being ready to embrace challenges.*
- *Suffering from imposter syndrome.*
- *Early expectation of success.*

In what follows, we detail each of the listed persona-related Ph.D. pitfalls.

6.2.1 Taking Everything Easy

Everything is hard before it is easy.

—Johann Wolfgang von Goethe (1749–1832)

Because earning a Ph.D. demands hard work and commitment for at least 3–5 years, you can easily be fatigued and stressed during the initial years of your Ph.D. study, when transitioning from being a graduate student to a research student. To manage fatigue and stress, your mind may consciously or unconsciously *take everything easy*[1] to the extent that it would feel great regularly, even without putting in the necessary hard work, experimentation, and thinking. This is precisely how the entitled persona-related Ph.D. pitfall can likely arise during the initial years of your Ph.D. study.

During the initial years of your Ph.D. study, your mind may take everything easy for many reasons. First, your academic performance during your master's study could perhaps be terrific enough to deceive your mind that you would also excel similarly during your Ph.D. study. Second, you may not know that the research world and its complicated global research enterprise are dynamic and unpredictable. Third, you may not certainly know what it takes to publish even a single top-tier journal paper until you succeed in publishing one. Fourth, you may not precisely know what it takes to be even a Ph.D. candidate. Last, you may not undoubtedly know what it takes—starting from *day one as a Ph.D. student*—to be a successful Ph.D. graduate. Accordingly, the entitled persona-related Ph.D. pitfall can likely happen during the initial years of your Ph.D. study.

Taking everything easy can be counterproductive and detrimental to your research progress during the initial years of your Ph.D. study. However, it can be a way of managing stress, fatigue, and negative emotions. To manage stress, fatigue, and negative emotions, you may need an optimal dosage of taking things easy, though taking things easy without any limit can be detrimental. Hence, there shall always be a sweet spot on the dosage of taking things easy, which you can infer through gigantic intellectual experience and realism about research, research careers, and research enterprises. Nevertheless, as a Ph.D. student and new researcher, you may not possess gigantic intellectual experience and realism about research, research career, and research enterprise. Therefore, the entitled persona-related Ph.D. pitfall boils down to multiple *irreducible chicken-and-egg problems*, which can confront you during the initial years of your Ph.D. study.

[1] If a Ph.D. (co-)supervisor observes a lack of research progress while you are taking everything easy, this is how and why you would get pushed—with a high probability.

6.2.2 Failing to Understand Yourself Very Well

Life consists in what a man is thinking of all day.

—Ralph Waldo Emerson (1803–1882)

Throughout your Ph.D. study, what is critical is the effective management of yourself, your Ph.D., and your Ph.D. (co-)supervisor, which you have to be also adept at during the initial years of your Ph.D. study, when you are expected to shoulder both coursework and research work simultaneously. While shouldering your Ph.D. responsibilities, you may sometimes go through emotional and daunting experiences. As a result, your attention can often be pulled to your stressful situation to the extent that you would hardly have time to be sufficiently self-aware and mindful of yourself going through different situations. This is how and why you may fail to understand yourself very well during the initial years of your Ph.D. study, justifying the likely onset of the entitled persona-related Ph.D. pitfall.

During the initial years of your Ph.D. study, you may likely fail to understand yourself—as a new researcher—very well for various reasons. First, while being possibly stressed by your scholarly situation and huge workload, you may fail to seize many opportunities to be mindful of yourself, though everything offers opportunities for being mindful of yourself. Second, you may not have or maintain the habit of self-awareness and self-reflection, especially in the heat of hard work and breathlessness. Third, since you may feel like you have no extra minute, you may hardly put adequate effort into completely understanding yourself as a new researcher, though time can always be properly managed for its availability to important things. Fourth, as everyone has weaknesses and strengths, you may also have weaknesses that you may fail to understand correctly, like many neuroscientifically subjective creatures, not fiercely objective ones. Fifth, you may not have the *time luxury* to meet and interact with kind professionals who may encourage your strengths and constructively criticize your weaknesses. Last, you may have never been put in a situation where you are a new researcher wrestling with the latest science and technology to get new publishable results. Consequently, the entitled persona-related Ph.D. pitfall can be inevitable throughout the initial years of your Ph.D. study.

If you fail to understand yourself, your mind sadly thinks in a world wherein you and yourself are not understood. As a result, what comes out of your mind—operating regarding its *misunderstood world*—can be counterproductive and harmful to your progress in the initial years of your Ph.D. study. Meanwhile, understanding yourself very well in the context of being a professional researcher entails remarkable self-awareness and consciousness of your mind while it goes through an enormous intellectual journey that is both challenging and rewarding. As a Ph.D. student and new researcher, however, your mind may not have yet gone through an enormous intellectual journey that is both challenging and rewarding. Therefore, the entitled persona-related Ph.D. pitfall comes down to many irreducible chicken-and-egg problems, which you may face throughout the initial years of your Ph.D. study.[2]

6.2.3 Not Being Ready to Embrace Challenges

Success is due to our stretching to the challenges of life. Failure comes when we shrink from them.

—John C. Maxwell (1979–)

As we frequently hear about those who achieve financial wealth of millions and billions quickly, this information age and twenty-first century seem all about shortcuts and fortnight successes. Contrary to the truth and in line with anecdotal evidence, such social media broadcasts unfortunately influence everyone following them, especially when they are heard repeatedly. From a *neuroscientific standpoint*, such repetitive stories can even produce a new *brain wiring* inside one's brain, completely altering one's *belief system* [1].[3] As a result, a drastic change in your belief system can make you—as a new researcher of the new generation—believe in only shortcuts, implicitly failing you in being ready to embrace challenges. This is why and how you may be confronted by the entitled persona-related Ph.D. pitfall throughout the initial years of your Ph.D. study.

Throughout your Ph.D. study, you may not be ready to embrace challenges due to many reasons. First, you may be subjective enough to think with a gambling mind that it would be great to get your Ph.D. done as soon as possible with minimum effort since the human brain can very well act with mental modes of wishful thinking and gambling. Second, you may not have sufficient research experience, intellectual confidence, and personal resilience to embrace challenges you would be determined to confront head-on. Third, you may not have a pragmatic experience in seizing challenges and transforming them into opportunities because your mind can be foreign to any challenge that can easily trigger its adverse reaction. Fourth, you may not have been positively influenced by a scholarly environment that often creates and seizes challenges as opportunities. Last, you may not have sufficient mental experience in creating and seizing challenges as significant personal and professional growth opportunities. Accordingly, you may fail to embrace challenges—in line with the mentioned change in your belief system—that can be a detrimental persona-related Ph.D. pitfall, which can dwarf

[2] In the worst-case scenario, you may also fail to understand yourself very well as a researcher during the middle and final years of your Ph.D. study.

[3] *Brain activity and structure* provide biological mechanisms for human behavior. Regarding human behavior, social networks provide external inducers/modulators [2, p. 34] (see also [3]). Often, social networks influence the development of our brains' *regional networks* through the developments of our *neuronal network*, which is influenced by the developments of our *cellular and molecular networks* [2, Box 2, p. 34].

your personal and professional growth throughout the initial years of your Ph.D. study.

Because a Ph.D. is awarded only to those who vividly demonstrate mastery of a research subject, research insight, respect for a discipline, capability for independent research, and ability to communicate scientific results and relate them to the broader discourse [4, p. 2], you shall always be ready to embrace challenges, which you can often seize as gigantic opportunities to satisfy these Ph.D. requirements via your prospective Ph.D. dissertation. Meanwhile, to successfully embrace challenges, you need to develop a sturdy mindset of embracing challenges while also creating them as opportunities, which you can achieve by going through numerous intellectual challenges that you would seize as ample opportunities. As a Ph.D. student and new researcher, you may not succeed in developing a sturdy mindset since you have not yet gone through many intellectual challenges that can inspire such a mindset. Therefore, the entitled persona-related Ph.D. pitfall comes down to various irreducible chicken-and-egg problems, which you may face throughout the initial years of your Ph.D. study.

6.2.4 Suffering from Imposter Syndrome

> *It's not what you are that holds you back, it's what you think you are not.*
>
> *—Denis Waitley (1933–)*

For any Ph.D. student and new researcher, the proof of every theorem you may come across can be too complicated to grasp easily; each system model and its mathematics can be too complex to comprehend effortlessly; every introduction may refer to several published papers that you have never read nor completely understood; and each experiment/simulation can be difficult to understand fully. While all these realities—among others—are frequent for many new Ph.D. students, your mind can doubtfully feel and believe that you are not as "intelligent and capable" as the previously published first authors who have elegantly contributed the "complicated journal papers" you come across. This is precisely how and why you may start suffering from *imposter syndrome (imposter phenomenon)* [5][4] throughout the initial years of your Ph.D. study.

Throughout the initial years of your Ph.D. study, you may suffer from imposter syndrome for numerous reasons. First, you may tend to be more conscious of what made you feel like an "imposter" than consciously reframing your thinking that you have become a Ph.D. student to learn, develop, grow, and innovate gradually. Second, you may have easily understood everything during your undergraduate and graduate studies to the extent that any of your struggles to understand journal

[4] Despite no rigorous and convincing evidence, your imposter syndrome can embody your subjective experience of self-doubting your abilities and experiences compared with other junior and senior authors of your research field [5].

papers feel abnormal to your mind. Third, you may not genuinely and openly discuss your genuine feelings of self-doubt with your Ph.D. (co-)supervisor and network of peers due to a fear of misunderstanding and insecurity. Fourth, you may be a *high-achieving and perfectionist* young researcher who sets ultra-high standards, which can be unrealistic expectations for you as a Ph.D. student and new researcher during your initial Ph.D. years. Fifth, you may not have adequate intellectual experience and expertise in navigating uncertain and unstructured research/innovation processes, which can easily trigger self-doubt. Sixth, the hyper-competitive academic culture and environment may force you to constantly prove your worth, easily triggering self-doubt and insecurity of being perceived as "less capable." Seventh, a doctoral study's usually high standards and expectations may force you to knowingly or unknowingly compare yourself with senior researchers and academics, instantly evoking feelings of inadequacy in your mind. Last, and most importantly, the mandatory and necessary learning curve of a Ph.D. study—such as developing diverse academic skills and mastering complex methodological approaches—can be tough enough to elicit feelings of self-doubt in your mind regularly.

The published first authors—whose papers you would come across—may have their paper rejected at first; may receive help and encouragement from several colleagues and collaborators; may be significantly led by their respective coauthors; and may have first struggled to solve the problems they have written their paper about. No wonder any part of their papers can be complicated for any Ph.D. student and new researcher such as yourself; hence, this shall not cause imposter syndrome. To overcome imposter syndrome, you need to neuroscientifically affirm that you are a worthy young researcher who can publish multiple top-tier journal papers despite their considerable challenges. However, as a Ph.D. student and new researcher, you have not yet published multiple top-tier journal papers. Therefore, the entitled persona-related Ph.D. pitfall comes down to several irreducible chicken-and-egg problems, which may confront you throughout the initial years of your Ph.D. study.[5]

6.2.5 Early Expectation of Success

Success is a slow process, and quitting does not speed it up.

—Jeffrey Fry

You may have been digging through many papers for months while giving up on your weekends, evening relaxation hours, social life, and spiritual times. While giving up on all these, you may have been tirelessly investigating various ideas and research directions. As a result, your mind may wind up engaging in wishful thinking while being emotionally attached to what you do, and you must be

[5] Imposter syndrome can also happen during the middle years of your Ph.D. study.

successful in your Ph.D. research as early as possible. During the initial years of your Ph.D. study, you may have been in this situation—that can also happen to any human as a neuroscientifically subjective creature—as a consequence of the entitled persona-related Ph.D. pitfall.

Throughout the initial years of your Ph.D. study, you may expect success early for many reasons. First, you may not know the persistent hard work needed to publish your first top-tier journal paper until you succeed in publishing one. Second, you may not have the pragmatic understanding and system thinking capability regarding a research career and enterprise, which is often the case for new researchers. Third, you may like to strongly believe that your persistent hard work would produce interesting research results, though any of your efforts can hardly guarantee any success in a research world that is always known to be uncertain. Fourth, because you may have witnessed others publishing multiple journal and conference papers in a short period, you may tend to think that you have to also be successful in a short period. Fifth, you may not be realistic enough about a research career to expect success early while overlooking the likelihood of failure in a world of research and experimentation. Last, you may have been getting early success during your pre-doctoral study so that you may expect the same for your Ph.D. study. Accordingly, the entitled persona-related Ph.D. pitfall can be inevitable throughout the initial years of your Ph.D. study.

Since novel research often needs not only the generation of novel ideas but also careful (and persistent), independent, and refinery thinking coupled with tireless endeavors of experimentation, success in research demands persistent efforts for an extended period while drawing numerous lessons from possibly wrong results. Meanwhile, to stop expecting success early, you may need to forge a pragmatic understanding of success in science by going through many uncertain intellectual journeys and achieving success while beating both the likelihood of failure and its consequent disappointment. Nevertheless, as a Ph.D. student and new researcher, you may not have yet gone through many uncertain intellectual journeys and attained success. Therefore, the entitled persona-related Ph.D. boils down to multiple irreducible chicken-and-egg problems, which you may face throughout the initial years of your Ph.D. study.

We now proceed with our detailed discussion of the persona-related Ph.D. pitfalls inevitable during the middle years of a Ph.D. study.

6.3 Persona-Related Pitfalls During the Middle Years of a Ph.D.

Our character is what we do when we think no one is looking.

—H. Jackson Brown, Jr. (1940–2021)

During the middle years of your Ph.D. study, you will also face various persona-related Ph.D. pitfalls, which are likely to be counterproductive and detrimental to your Ph.D. study. Some of these Ph.D. pitfalls are—but not limited to—itemized below.

! Attention: Persona-Related Pitfalls During the Middle Years of a Ph.D.

- *A lack of focus when things are not working.*
- *Drawing negative energy when things are not working.*
- *Getting distracted by wrong results.*
- *Following the crowd.*
- *Comparing yourself with others.*

In what follows, we discuss each of the itemized persona-related Ph.D. pitfalls.

6.3.1 A Lack of Focus When Things Are Not Working

Anything can happen, in spite of what you are pretty sure should happen.

—Richard P. Feynman (1918–1988)

During the middle years of your Ph.D. study, you may go through various unpleasant things such as getting stuck in your research, getting your conference/journal paper rejected, exhaustion, frustration of the hard work to be persisted daily, feeling disconnected and stuck in your personal life, and so forth. While all these things are regular occurrences of a Ph.D. routine that can happen to any Ph.D. student, your mind can have considerable adverse reactions by perceiving them as unusual events since you are a relatively new researcher. This is precisely how and why the entitled persona-related Ph.D. pitfall likely arises during the middle years of your Ph.D. study.

Whenever something you plan and work for is not working, you may immediately lack focus for several reasons. First, your Ph.D. realities may not match your research expectations. Second, your mind may not have gone through typical research processes for several years while successfully dealing with uncertainty and setbacks. Third, your mind may implicitly be figuring out the next steps while feeling the reality of crisis, negative energy, and uncertainty. Fourth, the things that happen not to work can easily pull your attention from what you have been focusing on because your mind can resort to an emergency mode. Last, those things can also evoke immense negativity in your head, effectively destroying your focus for many days—if not weeks. As a result, the entitled persona-related Ph.D. pitfall is unavoidable during the middle years of your Ph.D. study.

When things are not working, you may lose focus on nearly everything you hope to accomplish, which you may have been planning for quite some time. Meanwhile, to lose a minimum or no focus on your research, irrespective of whether things are working or not, you need to have considerable research experience while successfully managing its inherent uncertainty and challenges. As a Ph.D. student and relatively new researcher, however, you have not yet garnered significant research experience by managing the inherent uncertainty and challenges of scientific research. Therefore, the entitled persona-related Ph.D. pitfall comes down to many irreducible chicken-and-egg problems, which you may face during the middle years of your Ph.D. study.

6.3.2 Drawing Negative Energy When Things Are Not Working

Negativity is the enemy of creativity.

—David Lynch (1946–2025)

In an intellectual and research world of experimentation such as your Ph.D. study, several things may either go wrong or hardly work at first because your Ph.D. study would chiefly be a time of experimentation and uncertainty, so anything expected or desired can fail to materialize anytime. To this end, there may be many objective reasons why things are not working, which you may not have figured out through experimentation and thinking. While things may not commonly work as planned during anyone's Ph.D. study, your mind may draw negative energy quickly. This is precisely how and why the entitled persona-related Ph.D. pitfall probably happens during the middle years of your Ph.D. study.

During the middle years of your Ph.D. study, you will probably draw negative energy, which will be a counterproductive force against your focus and determination for many reasons. First, your optimistic expectations may not match the bitter realities of your Ph.D. study and journey. Second, your Ph.D. realities can sharply and disappointingly contradict your research expectations, forcing you to draw immense negative energy. Third, your mind may knowingly or unknowingly engage in wishful thinking, suppressing the necessary rigorous thinking before any experimentation/simulation. Fourth, like any human, you are a neuroscientifically subjective creature with emotions, likes, dislikes, preferences, biases, etc. Last, in the heat of your Ph.D. pressure and its funding constraints, you may often rush to the extent that you do not have time to be thoughtful. Consequently, the entitled persona-related Ph.D. pitfall is inevitable during the middle years of your Ph.D. study.

To not emotionally draw negative energy whenever things are not working, you may need to develop a strong mind, which can seize every challenge as an opportunity through tons of research experience, pragmatism, and wisdom. Nevertheless, as a Ph.D. student and relatively new researcher, your research experience, pragmatism, and wisdom—among others—may not yet be substantial. Therefore, the entitled persona-related Ph.D. pitfall boils down to various irreducible chicken-and-egg problems, which you may face during the middle years of your Ph.D. study.

6.3.3 Getting Distracted by Wrong Results

When things do not go your way, remember that every challenge – every adversity – contains within it the seeds of opportunity and growth.

—*Roy T. Bennett (1963–)*

After believing you have a great idea worth investigating and publishing, you may work tirelessly on the investigation for several months and obtain frustratingly wrong results. Because things may not work as you hope, you may abruptly decide to wander around—physically and neuroscientifically—and spend too much of your valuable time for nothing. You may then get anxious and unsuspectingly irrational that you will flunk out of your Ph.D. program since you get only the wrong results. This is how and why you may get distracted by the wrong results for much longer than necessary, hence the viability of the entitled persona-related Ph.D. pitfall.

During the middle years of your Ph.D. study, you may likely be distracted by wrong results—popping up on different occasions—for the following reasons. First, wrong results—during a busy Ph.D. research season—can cause anxiety, depression, stress, and immense negativity. Second, you may not know that wrong results are inevitable in the life of any researcher since research is all about experimentation and thinking. Third, in the heat of the wrong-result moment, you may likely fail to see the *silver lining* of the wrong results, which are gigantic opportunities for further research, thinking, and experimentation. Fourth, wrong results can easily demotivate any relatively new researcher such as yourself. Fifth, wrong results can cause immense distraction and despair, especially for new researchers. Last, wrong results can likely deceive many minds of new researchers that the entire effort was a mere waste of time, so one has to start again from scratch. Consequently, the entitled persona-related Ph.D. pitfall is inevitable during the middle years of your Ph.D. study.

Since your success hinges on the effective utilization of your scarce time and attention regularly, how much and how long you would be distracted by wrong results may determine your success. Meanwhile, not to be distracted by wrong results for long, you need to develop a strong, swift, adaptive, and goal-oriented mindset—that can move on anytime—via gigantic research experience, pragmatism, and expertise. As a Ph.D. student and relatively new researcher, nevertheless, you have not yet succeeded in—among others—garnering gigantic research experience and pragmatism along with developing considerable subject-matter expertise. Therefore, the entitled person-related Ph.D. pitfall boils down to many irreducible chicken-and-egg problems, which can confront you throughout the middle years of your Ph.D. study.

6.3.4 Following the Crowd

You cannot dream yourself into a character; you must hammer and forge yourself one.

—Henry David Thoreau (1817–1862)

The technologies of the twenty-first century can knowingly or unknowingly make you follow the crowd in many ways. For instance, you may simply follow many scientists and non-scientists on social media; you may follow papers with huge citation numbers; you may believe and follow researchers with a huge *h-index*; you may follow researchers with enormous collaborators; and so on. On the other hand, you may follow researchers who frequently chase the information/research trend—without any depth and breadth of fundamental endeavor—for the sake of citation. This is precisely how astronomically many researchers pursue *science in the age of selfies* [6] and how you can be detrimentally impacted by the entitled persona-related Ph.D. pitfall throughout the middle years of your Ph.D. study.

You may follow the crowd during the middle years of your Ph.D. study for several reasons. First, your mind can operate in a gambling mode that following the crowd entails much less risk. Second, co-existing and moving with the crowd may make many new researchers, such as yourself, feel it is a collective way to minimize risk. Third, you may witness numerous Ph.D. students in your research field publishing in not-so-renowned journals, making you target the same journals for their high acceptance rate—at least when you are frustrated. Fourth, you may observe many Ph.D. students in your research field adopting a widely known or easy technique/methodology, enticing you to do the same. Fifth, your mind can operate in a gambling mode to strongly believe that following the trend like the crowd is the easiest way out. Last, you may lack significant research experience, expertise, and confidence to lead wisely while ignoring the crowd. As a result, the entitled persona-related Ph.D. pitfall is unavoidable throughout the middle years of your Ph.D. study.

While you may need to be a follower to some degree as a new researcher looking to carve out one's independent scientific path, being simply a follower is a *suicide mission* of a thoughtless (or vision-less) researcher who may not be blessed with courage since the overarching goal of a great researcher is being a (thought) leader—never a follower. Meanwhile, to be a thought leader, you may need to accomplish extraordinary science, technology, engineering, and mathematics (STEM) through remarkably independent thinking of depth and breadth, stellar vision, gigantic courage, and proactivity. As a Ph.D. student and relatively new researcher, however, you may not be equipped with—among others—the skill set for remarkably independent thinking of depth and breadth, stellar vision, and proactivity. Therefore, the entitled persona-related Ph.D. pitfall comes down to numerous irreducible chicken-and-egg problems, which you would be confronted by throughout the middle years of your Ph.D. study.

6.3.5 Comparing Yourself with Others

In the absence of a great dream, pettiness prevails.

—Robert Fritz (1943–)

Academia and the Ph.D. system are always hyper-competitive and intellectually demanding by nature, to the degree that many Ph.D. students end up comparing themselves with others almost daily. The comparison with others is further aggravated by the fact that Ph.D. students are continually assessed through journal publications, conference presentations, peer reviews, and internal scholarship applications. This is how and why the entitled persona-related Ph.D. pitfall arises throughout the middle years of your Ph.D. study.

Throughout the middle years of your Ph.D. study, you may frequently compare yourself with others for numerous reasons. First, comparing yourself with others is how you gauge your research performance as a relatively new researcher. Second, your Ph.D. (co-)supervisor may tend to compare you with others in one's lab or outside. Third, your mind may be looking for a reason to engage in wishful thinking, such that you are doing much better than those who do not publish regularly. Fourth, your university's competitive and scarce internal scholarship forces you to compare with others. Fifth, you may not consciously realize the amount of valuable time you often burn while comparing yourself with others. Sixth, you may be advised to compete with others, making you compare yourself regularly. Last, you may not know how to rigorously develop yourself into a rising star without comparing yourself with others. Consequently, the entitled persona-related Ph.D. pitfall is inevitable throughout the middle years of your Ph.D. study.

Regarding your unique *inherent neuronal system* and *belief system*, you can succeed only in your way—not in others' way or belief system—though the lessons you would gain from others' success can be helpful in your success strategy. Your formidable challenge is thus carving out your unique way while being adaptive, well-informed, flexible, wise, and pragmatic. Meanwhile, to not compare yourself with others and be driven by their successes, you may need to mature into a *principle-centered researcher and thinker* who is forged by considerable research experience and pragmatism, unflinching values, and great purpose in STEM. Nonetheless, as a Ph.D. student and relatively new researcher, you may not possess—among others—considerable research experience and pragmatism to forge yourself into the mentioned person. Therefore, the entitled persona-related Ph.D. pitfall boils down to various irreducible chicken-and-egg problems, which you would face throughout the middle years of your Ph.D. study.

We now continue with the motivation-related Ph.D. pitfalls that are inevitable during the initial years of your Ph.D. study.

6.4 Motivation-Related Pitfalls During the Initial Years of a Ph.D.

You will become large as your controlling desire; or as great as your dominant aspiration.

—James Allen (1864–1912)

During the initial years of your Ph.D. study, you will likely be confronted by various motivation-related Ph.D. pitfalls, which can significantly harm your Ph.D. research progress. Some of these Ph.D. pitfalls are—but not limited to—listed below.

! Attention: Motivation-Related Pitfalls During the Initial Years of a Ph.D.

- *Being overwhelmed.*
- *Working long hours alone.*
- *A lack of momentum.*
- *A lack of motivation.*
- *A lack of positive energy and drive.*

Next, each of the listed motivation-related Ph.D. pitfalls is discussed separately.

6.4.1 Being Overwhelmed

You are the sky. Everything else – it's just the weather.

—Pema Chödrön (1936–)

During the initial years of your Ph.D. study, you may come across the most complex paper you have ever seen, the longest paper you have ever seen, and the most complex mathematics and proofs you have ever seen. As you keep reading many other papers, you may have the same impression, huh, to your surprise. You may then be stunned by the "intelligence" of all authors for managing and delivering such "incredible papers" to your research field. While these are very normal feelings among new researchers entering new research fields, your mind may be overwhelmed throughout the initial years of your Ph.D. study. This is how and why the entitled motivation-related Ph.D. pitfall arises throughout the mentioned time.

You can frequently be overwhelmed throughout the middle years of your Ph.D. study for many reasons: first, a lack of confidence and self-esteem as a new researcher since you would understandably be new to your research field and scientific community; second, insufficient prior research and intellectual experience in your research field, if any; third, a lack of diverse scholarly experience and contributions to your research field; fourth, a lack of focus as you may end up touching many things in the hope of mastering the trends of your research field and making contributions; fifth, you may lack a depth and breadth of subject-matter expertise in your research field; sixth, you may not know where to focus because you may also lack subject-matter expertise in your research field; seventh, the usually complex works and state-of-the-art advancements of your research field, considering that you would be just a beginner in your research field; last, being overwhelmed is generally the natural reaction of any mind that comes across a totally novel and seemingly challenging thing. Consequently, the entitled motivation-related Ph.D. pitfall is unavoidable throughout the initial years of your Ph.D. study.

Not only you but also any Ph.D. student and new researcher can be overwhelmed by anything new and challenging that you may regularly come across since that would be the natural reaction of anyone's mind dealing with a novel situation it may perceive as challenging. To not be overwhelmed, you must deliver multiple contributions and develop advanced expertise in your research field. However, as a Ph.D. student and new researcher, you may not have made several contributions or developed advanced expertise in your research field. Therefore, the entitled motivation-related Ph.D. pitfall comes down to several irreducible chicken-and-egg problems, which you may face throughout the initial years of your Ph.D. study.

6.4.2 Working Long Hours Alone

> *Success is not accident. It is hard work, perseverance, learning, studying, sacrifice and most of all, love of what you are doing or learning to do.*
>
> *—Pelé (1940–2022)*

Because you would be spending much of your time in uncharted scholarly territories, a (novel) Ph.D. study usually needs substantial hard work and determination from the beginning to the end since it is always intellectually demanding. The many inevitable Ph.D. pitfalls also force you to spend long hours almost daily, especially during the initial years of your Ph.D. study. During this transition period that can test your stamina to the fullest, you are required to demonstrate that you can handle both coursework and research work effectively. Accordingly, you may work long hours alone throughout the initial years of your Ph.D. study because working tirelessly hard may be the only way out.

During the initial years of your Ph.D. study, you may work long hours alone for many reasons. First, the high expectations of your Ph.D. (co-)supervisor and graduate school so that you could satisfy the minimum requirements to be a Ph.D. candidate. Second, the high intellectual standards of a Ph.D. research, at least in comparison with your previous research experience. Third, a lack of research experience and expertise in your research field. Fourth, you may hardly have foundational research work to build on. Fifth, you should conduct your Ph.D. research independently under your Ph.D. (co-)supervisor's supervision/guidance. Sixth, you have not yet mastered your research field's state-of-the-art knowledge and advancements. Seventh, there is usually intensive competition in your research discipline and field to get published as much as possible while making the most plausible impact. Last, you may have limited or no professional writing and publication experience thus far. Consequently, the entitled motivation-related Ph.D. pitfall is inevitable throughout the middle years of your Ph.D. study.[6]

As you depend less on your Ph.D. (co-)supervisor, you depend more on long working hours to grasp everything and figure out how to proceed best—all by yourself. This is how you can demonstrate your intellectual independence throughout your Ph.D. study while effectively working long hours alone. Meanwhile, to effectively work productive working hours, you may need to develop—among others—considerable expertise in your research field so that you can easily understand how things go and where they can be interconnected in a novel way. However, as a Ph.D. student and new researcher, you may not yet have developed your research field's significant expertise, which can enable you to innovate on the go. Therefore, the entitled motivation-related Ph.D. pitfall boils down to several irreducible chicken-and-egg problems, which you may come across throughout the initial years of your Ph.D. study.

6.4.3 A Lack of Momentum

It does not matter how slowly you go as long as you do not stop.

—Confucius (ca. 551 - ca. 479 BC)

During the initial years of your Ph.D. study, your Ph.D. (co-)supervisor may have high expectations for your coursework and research work. What is quite challenging is when such high expectations are compounded with your lack of time management skills, shortage of research experience, and absence of previous research works to build upon. Your quite challenging situation can even be worsened by the many

[6] As we will discuss later, this motivation-related Ph.D. pitfall is also inevitable throughout the middle years of your Ph.D. study.

diverse Ph.D. pitfalls you may encounter throughout the initial years of your Ph.D. study when you are expected to manage both coursework and research work concurrently, which could spell further challenges along the road to the middle years of your Ph.D. study. Consequently, you may struggle from a lack of momentum throughout the initial years of your Ph.D. study.

During the initial years of your Ph.D. study, you may lack momentum for several reasons. First, you may not have prior publications that can provide a foundation and momentum for your upcoming works. Second, your Ph.D. research project can be quite different from your master's thesis research, causing you a lack of momentum unless you pursue a master's plus Ph.D. study. Third, due to the availability of astronomically many state-of-the-art research works, due to the rise of low-impact publication venues, you may touch here and there rather than being laser focused on your research problem, causing you a lack of momentum as you keep getting distracted. Fourth, the presence of abundant distracting literature can cause you a lack of momentum since each distracting literature can take your essential focus away from your specific research problem. Fifth, the many diverse Ph.D. pitfalls inevitable throughout your Ph.D. study can considerably demotivate you, sapping your momentum during the initial years of your Ph.D. study. Sixth, because you have to demonstrate your intellectual independence throughout your Ph.D. study, you may not be allowed to work with potential collaborators whose research experience, subject-matter expertise, and guidance would otherwise be momentum for your research progress. Last, you may not know precisely how to start creatively small—concerning your huge research problem—to quickly produce a publication, which would give you momentum to progress further. As a result, the entitled motivation-related Ph.D. pitfall is inevitable throughout the initial years of your Ph.D. study.

To get momentum that can boost your further research progress, you may need to produce—among others—several important works that you can build on; however, you need momentum to successfully produce various important works that you can build on further. Therefore, the entitled motivation-related Ph.D. pitfall boils down many irreducible chicken-and-egg problems, which you may face throughout the initial years of your Ph.D. study.

6.4.4 A Lack of Motivation

People often say that motivation doesn't last. Well, neither does bathing – that's why we recommend it daily.

—Zig Ziglar (1926–2012)

During the initial years of your Ph.D. study, you will encounter many diverse Ph.D. pitfalls, such as research-related, writing and publication-related, management-related, and persona and motivation-related Ph.D. pitfalls. Regarding the seemingly never-ending hard work to persist throughout the initial years of your Ph.D. study, these pitfalls, compounded by a lack of momentum, can be a recipe for a *magnificent lack of motivation*. This is how and why the entitled motivation-related Ph.D. pitfall can likely arise throughout the initial years of your Ph.D. study.

During the initial years of your Ph.D. study, your mind can manifest a lack of motivation for many reasons. First, your mind may not practically know that some wrong results—often perceived as "failures"—are needed to progress in your research. Second, you may often be negative and somehow stressed about your prospects in a research career, considering the initial Ph.D. years' challenges you may go through that your mind may perceive as hardship. Third, regarding the significant challenges of managing coursework and research work simultaneously, you may lose your self-confidence and self-belief that you can make important contributions to STEM. Fourth, you may lack ambition, persistence, and focused effort, considering the challenging situation you may be in. Fifth, you may have a wrong attitude about a Ph.D. and research, continually dwindling your original motivation for a Ph.D. study. Sixth, you may knowingly or unknowingly embed yourself in the wrong environment, which can consistently challenge your inherent motivation for a Ph.D. study and research career. Seventh, you can be immensely impacted by your presence on social media, which inadvertently forces you to compare your "hard life" with others' advertised "joyous lives." Last, you may follow the crowd who may knowingly or unknowingly demotivate you amid your initial Ph.D. years of challenges. As a result, the entitled motivation-related Ph.D. pitfall is unavoidable throughout the initial years of your Ph.D. study.

The absence of motivation would probably be one of the most crippling motivation-related Ph.D. pitfalls that you may suffer from throughout the initial years of your Ph.D. study. Meanwhile, persistent and productive hard work is carried out by a mind motivated by adequate research momentum and promising research results; on the contrary, adequate research momentum and promising research results are produced by a motivated mind.[7] Therefore, the entitled motivation-related Ph.D. pitfall comes down to numerous irreducible chicken-and-egg problems, which can trouble you throughout the initial years of your Ph.D. study.

6.4.5 A Lack of Positive Energy and Drive

Success is the sum of small efforts – repeated day in and day out.

—Robert Collier (1885–1950)

During the initial years of your Ph.D. study, you can be regularly overwhelmed; you may end up working long hours alone; you may struggle from a lack of momentum;

[7] You would also face a motivation-related Ph.D. pitfall due to a lack of motivation during the middle years of your Ph.D. study, as discussed subsequently.

and you may suffer from a drastic lack of motivation. These bitter realities of being a Ph.D. student and new researcher can easily cause you a lack of positive energy and drive throughout the mentioned period, asserting the high likelihood of facing the entitled motivation-related Ph.D. pitfall.

Throughout the initial years of your Ph.D. study, your mind can exhibit a lack of positive energy and drive due to various reasons. First, you may have self-doubt about your capabilities as a new researcher with little to no scientific contributions. Second, the many inevitable Ph.D. pitfalls can make you pessimistic, exhausted, and hopeless, sapping your positive energy and drive regularly. Third, whatever you may come across can feel challenging and scary, and so can many Ph.D. students and new researchers. Fourth, the uncertain and intellectually challenging nature of a new researcher's scientific journey, regularly depleting your positive energy and drive. Fifth, because research is also an emotional experience, especially while working under deadlines as a new researcher, anything that may fail to work can easily cause your mind negativity and a lack of drive. Sixth, you may unreasonably feel that you are the only one going through the many challenges you encounter, though others may also endure similar challenges. Seventh, your realities can be drastically different from your expectations since your mind can wishfully expect early success. Last, you may feel that what you get into is not the pursuit of boundless fascination and curiosity but an endless struggle and breathlessness. As a result, the entitled motivation-related Ph.D. pitfall is unavoidable during the initial years of your Ph.D. study.

To be a successful researcher, you need—among others—to have magnificent positive energy and drive; on the contrary, you need to be a successful researcher to have magnificent positive energy and drive. Therefore, the entitled motivation-related Ph.D. pitfall boils down to several irreducible chicken-and-egg problems, which may confront you throughout the initial years of your Ph.D. study.

We now proceed with the motivation-related Ph.D. pitfalls unavoidable during the middle years of your Ph.D. study.

6.5 Motivation-Related Pitfalls During the Middle Years of a Ph.D.

Personal development is a major time-saver. The better you become, the less time it takes you to achieve your goals.

—Brian Tracy (1944–)

The middle years of your Ph.D. study would often be ridden with various research-related, writing and publication-related, management-related, and persona-related Ph.D. pitfalls. While working hard to overcome these pitfalls, attempting to move on with your Ph.D. research usually triggers an internal crisis: *a lack of inherent motivation*. Throughout the middle years of your Ph.D. study, a lack of motivation coupled with seemingly never-ending hard work evokes troublesome motivation-related Ph.D. pitfalls. Some of these Ph.D. pitfalls are—but not limited to—listed below.

! Attention: Motivation-Related Pitfalls During the Middle Years of a Ph.D.

- *Working long hours alone.*
- *Staying focused and motivated when rejected.*
- *Struggling to stay motivated.*
- *Frustration with the Ph.D. process.*
- *A feeling of exhaustion.*

Next, we discuss each of the itemized motivation-related Ph.D. pitfalls.

6.5.1 Working Long Hours Alone

Tough times never last, but tough people do.

—Robert Schuller (1926–2015)

What you must go through as a Ph.D. student—such as the usually demanding Ph.D. requirements, high intellectual standards, scrutiny by your Ph.D. (co-)supervisor, and tough peer reviews—often translates into astronomical hours that you have to put into your Ph.D. research. Consequently, you may likely be overwhelmed and work long hours regularly throughout the middle years of your Ph.D. study, when you are especially expected to deliver several novel journal papers. This is how and why the entitled motivation-related Ph.D. pitfall arises during the middle years of your Ph.D. study, besides the initial years of your Ph.D. study.

During the middle years of your Ph.D. study, you may likely work long hours alone for many reasons. First, you need to conduct several research investigations to succeed in publishing quality journal papers. Second, you may lack depth and breadth of research experience and expertise in your research field. Third, the high intellectual standards of many top-tier journals translate to numerous work hours every week. Fourth, the high expectations of your Ph.D. (co-)supervisor and graduate school so that you can satisfy the minimum requirements for a Ph.D. graduation. Fifth, you may not have a seminal or high-impact work to build on. Sixth, you may have limited or no advanced professional writing and publication experience thus far. Last, intense competition—not only in academia but also in your research field—to get extensively published while making a significant impact. As a result, the entitled motivation-related Ph.D. pitfall is inevitable during the middle years of your Ph.D. study.

Not to work long hours alone while possibly being overwhelmed, you need to be in control after producing several novel results; on the contrary, you need to work long hours alone while possibly being overwhelmed to produce several novel results. Therefore, the entitled motivation-related Ph.D. pitfall comes down to many irreducible chicken-and-egg problems, which you may face during the middle years of your Ph.D. study.

6.5.2 Staying Focused and Motivated When Rejected

> *If you're walking down the right path and you're willing to keep walking, eventually you'll make progress.*
>
> —*Barack Obama (1961–)*

Ideally, any paper could be scrutinized and rejected concerning *infinite depth and breadth of expertise*. For instance, a paper that contributes immensely in depth can be further scrutinized in depth and immense breadth, leading to its rejection. Similarly, any paper that contributes significantly in breadth can be further scrutinized in breadth and magnificent depth, inspiring its immediate rejection. Hence, because any paper can be rejected regardless of its contributions, many researchers unanimously agree that *rejection is part of any researcher's life,*[8] so it is part of your life as a relatively new researcher. To this end, staying focused and motivated when rejected would undoubtedly be challenging throughout the middle years of your Ph.D. study, hence the entitled motivation-related Ph.D. pitfall.

During the middle years of your Ph.D. study, you may struggle to stay focused whenever rejected for many reasons. First, rejection can temporarily cause loss of focus, pessimism, demotivation, frustration, and exhaustion, which can last longer as you lose motivation and focus. Second, rejection can be bitter, demotivating, and discouraging for any Ph.D. student and relatively new researcher since the rejected results are often obtained after several months, if not years, of hard work. Third, rejection can cause confusion and uncertainty on the intellectual path you have been pursuing. Fourth, rejection can cause a lack of self-confidence to the extent that you lose your original motivation and become clueless about where to focus next. Fifth, you may not have ample intellectual experience to back up when "failing" with temporary setbacks such as rejection. Last, you may feel defeated entirely for much longer, especially when the rejection happens after your considerable struggle to deliver meaningful results. Hence, the entitled motivation-related Ph.D. pitfall is unavoidable during the middle years of your Ph.D. study.

Top-tier journals and conferences are known for their low paper acceptance rate and high rate of rejection. Rejection is also a means of quality control for journals that tend to target a high-impact factor and wider readership, making rejection

[8] These arguments are aimed at promoting *free thinking, deep thinking, and broad thinking* that may inspire motivation and alternative perspectives. However, they shall not be exploited to belittle or disrespect the contributions of many leading researchers.

inevitable. Meanwhile, to stay focused and motivated when rejected, you must have sufficient rejection experiences that you have managed swiftly and successfully as ample research opportunities. As a Ph.D. student and relatively new researcher, however, you may not have sufficient rejection experiences that you have managed swiftly and successfully as ample opportunities. Therefore, the entitled motivation-related Ph.D. pitfall boils down to many irreducible chicken-and-egg problems, which can trouble you throughout the middle years of your Ph.D. study.

6.5.3 Struggling to Stay Motivated

When you do things from your soul, you feel a river moving in you, a joy.

—Rumi (1207–1273)

As a Ph.D. student and relatively new researcher, managing yourself, your Ph.D. (co-)supervisor, and your Ph.D.—its research quality, writing standard, and publication prestige—can be pretty challenging throughout the middle years of your Ph.D. study, when you are particularly expected to deliver top-tier journal papers. Concerning the inevitability of paper rejections, the feelings of exhaustion from persistent hard work, uncertainty, and *being in the middle of the tunnel*; staying motivated throughout the middle years of your Ph.D. study; and satisfying high intellectual standards can be considerable challenges. This is how and why the entitled motivation-related Ph.D. pitfall happens during the middle years of your Ph.D. study.

Throughout the middle years of your Ph.D. study, you may struggle to stay motivated for various reasons. First, despite your years of endeavors and commitment, you may not obtain as many novel results as you want. Second, you may be fed up with the weekly—if not daily—commitment you have been through for multiple years. Third, you may be exhausted by your Ph.D. study's insatiable need for hard work and excellence throughout the years. Fourth, considering your hard work for multiple years, you may not be satisfied with your achievements in terms of published top-notch journal papers. Fifth, you may be exhausted by your Ph.D. (co-)supervisor's continual demand for new results and papers. Sixth, you may be working on a nearly saturated and old research field to which you have struggled to contribute. Seventh, you may battle with a challenging and new research field that needs almost a daily struggle and full attention. Last, you may often yearn for success and the end of the tunnel. Accordingly, the entitled motivation-related Ph.D. pitfall is inevitable during the middle years of your Ph.D. study.

To succeed as a Ph.D. student and relatively new researcher, you must be motivated throughout the middle years of your Ph.D. study. To be motivated throughout the middle years of your Ph.D. study, you must succeed as a Ph.D. student and relatively new researcher. Therefore, the entitled motivation-related

Ph.D. pitfall comes down to many irreducible chicken-and-egg problems, which you can face during the middle years of your Ph.D. study.

6.5.4 Frustration with the Ph.D. Process

> *To conquer frustration, one must remain intensely focused on the outcome, not the obstacles.*
>
> *—T.F. Hodge (1969–)*

Throughout your Ph.D. study, you are expected to work under the leadership of your Ph.D. (co-)supervisor to produce, revise, edit, and submit journal and conference papers. Your submitted papers would then have to go through peer review at their respective conferences and journals. In the case of your journal submission, the peer reviewers may send you numerous comments and suggestions that you must address and incorporate, respectively, before responding with your reviewed manuscripts. Another round of major, minor, or rejection revisions may follow. If you succeed during the second round of review, your journal paper will be accepted and published, resetting the entire process for your upcoming submission. This is how and why you may be frustrated with the Ph.D. process during the middle years of your Ph.D. study, when you are specifically expected to deliver multiple novel journal papers.

During the middle years of your Ph.D. study, you may be frustrated with the Ph.D. process for various reasons. First, you may be exhausted from the demanding cycle of journal paper submission, revision, acceptance, and then submission. Second, you can often be overwhelmed by the unpredictably uncertain nature of the peer review process. Third, satisfying the high intellectual standards of the multi-round peer review is burdening. Fourth, each journal paper submission and revision round can feel like an endless struggle for a high intellectual standard. Last, should you disagree with your Ph.D. (co-)supervisor on when and where to submit, you will likely be more frustrated and pessimistic about what you would go through. Consequently, the entitled motivation-related Ph.D. pitfall can recur throughout the middle years of your Ph.D. study.

You shall often remind yourself that there is always a way out, regardless of the seemingly worse situation you could be in. Meanwhile, you may need to be frustrated first to be a successful Ph.D. student and relatively new researcher. To be a successful Ph.D. student and relatively new researcher, on the other hand, you shall always be ready to tirelessly shoulder your research responsibilities for a high intellectual standard without being frustrated. Therefore, the entitled motivation-related Ph.D. pitfall boils down to many irreducible chicken-and-egg problems, which can trouble you during the middle years of your Ph.D. study.

6.5.5 Feelings of Exhaustion

I had many moments of disappointment, despondency, and exhaustion, but I always found that by reading the literature and showing up at my lab looking at the data as they emerged day by day and discussing them with my students and postdoctoral fellows, I would gain a notion of what to do next.

—Eric Kandel (1929–)

In a Ph.D. study, everything demands hard work and persistence from the beginning till the end. First, the necessary hard work starts with a steep learning curve to master the many state-of-the-art developments of your research field. Second, it revolves around innovation through new idea generation, problem formulation, proposing solutions, and computer experiments and/or simulations. Third, it would revolve around paper writing, editing, and submission. Fourth, it would then revolve around engaging with the possibly multiple rounds of peer review to get your paper published. You would then need to go through steps two to four to publish another paper and so forth. As a result, these demanding intellectual standards would make you feel exhausted throughout the middle years of your Ph.D. study.

Throughout the middle years of your Ph.D. study, you can feel exhausted since you must stay committed to qualifying for high research, innovation, writing, and publication standards while staying committed to satisfy the challenging demands of anonymous peer reviewers. As a result, the entitled motivation-related Ph.D. pitfall is unavoidable during the middle years of your Ph.D. study.

You have to be exhausted first in order to be successful. To be successful, on the other hand, you should not be exhausted. Therefore, the entitled motivation-related Ph.D. pitfall comes down to numerous irreducible chicken-and-egg problems, which can confront you throughout the middle years of your Ph.D. study.

6.6 Concluding Remarks

The only limit to our realization of tomorrow will be our doubts of today.

—Franklin D. Roosevelt (1882–1945)

This chapter presented various motivation and persona-related Ph.D. pitfalls unavoidable during the initial and middle years of your Ph.D. study. Concerning the initial years of your Ph.D. study, we detailed persona-related Ph.D. pitfalls such as taking everything easy, failing to understand yourself very well, not being ready to embrace challenges, suffering from imposter syndrome, and early expectation of success. Concerning the initial years of your Ph.D. study, we also detailed motivation-related Ph.D. pitfalls such as being overwhelmed, working long hours alone, a lack of momentum, a lack of motivation, and a lack of positive energy and drive.

Regarding the middle years of your Ph.D. study, we discussed persona-related Ph.D. pitfalls such as a lack of focus when things are not working, drawing negative energy when things are not working, getting distracted by wrong results, following the crowd, and comparing yourself with others. We also discussed motivation-related Ph.D. pitfalls such as working long hours alone, staying focused and motivated when rejected, struggling to stay motivated, frustration with the Ph.D. process, and feelings of exhaustion.

Even though we tried to discuss the most important motivation and persona-related Ph.D. pitfalls unavoidable during the initial and middle years of your Ph.D. study, our limited discussion—due to a space constraint—may not have covered the *full-blown depth and breadth* of issues interwoven with the discussed themes. For detailed insights and realities, we encourage you to reflect this chapter's references and discuss them with your network of peers.

References

1. Lipton BH (2008) The Biology of Belief: Unleashing the Power of Consciousness, Matter, & Miracles, Hay House, Carlsbad, CA, USA
2. Lynn CW, Bassett DS (2019) The Physics of brain network structure, function, and control. ArXiv. https://arxiv.org/pdf/1809.06441.pdf
3. Lynn CW, Bassett DS (2019) The Physics of brain network structure, function, and control. Nat. Rev. Phys. 1:318–332
4. Petre M, Rugg G (2010) The Unwritten Rules of PhD Research, 2nd edn. Open University Press, Berkshire, UK
5. Wikipedia (2024) Imposter syndrome. https://en.wikipedia.org/wiki/Imposter_syndrome. Cited Dec. 2024
6. Geman D, Geman S (2016) Science in the age of selfies. Proc. Natl. Acad. Sci. (PNAS) 113:9384–9387

Chapter 7
Generic Pitfalls of a Ph.D.

Don't be satisfied with stories, how things have gone with others. Unfold your own myth.

—Rumi (1207–1273)

Abstract The intellectually demanding nature of a Ph.D. study and its many uncertainties give rise to numerous specific Ph.D. pitfalls. To overcome the cumulative impact of these pitfalls, you need to develop as a critical thinker, a creative thinker, a strategic thinker, a probabilistic thinker, a causal thinker, and an independent thinker by going through a routine developmental process, including "failures." This leads to numerous irreducible chicken-and-egg problems that effect generic Ph.D. pitfalls throughout your Ph.D. study. Accordingly, this chapter discusses the generic Ph.D. pitfalls of a Ph.D. so that you would be inspired to prepare yourself in advance.

Keywords Generic Ph.D. pitfalls · Critical thinking · Creative thinking · Strategic thinking · Probabilistic thinking · Causal thinking · Independent thinking

7.1 Introduction

There are the things that are out in the open, and there are the things that are hidden. The real world has more to do with what is hidden.

—Saul Leiter (1923–2013)

Chapters 3–6 discussed various specific Ph.D. pitfalls, such as research-related, writing and publication-related, management-related, and persona and motivation-related Ph.D. pitfalls, respectively. Inevitably, during the different phases of your Ph.D. study, many of these Ph.D. pitfalls are often encountered simultaneously in the *realistic and non-linear world* of being a Ph.D. student. As a result, maximizing the probability of success in your Ph.D. study has never been a straightforward endeavor

T. M. Getu, *Pathways for Inspired Scientists and Engineers*,
https://doi.org/10.1007/978-3-032-06784-5_7

that paves the way to overcome the specific Ph.D. pitfalls whose cumulative impact can, unfortunately, be cascaded in a non-linear manner.

As for non-linearity, the realistic world is often the union of many small or larger events that commonly have their intricacies and realities. In this regard, everything naturally constitutes an interconnected system—directly or indirectly—so that minor hidden change in one thing often results in a *ripple effect* across the entire system. Following this school of thought that arguably captures the rapidly and unpredictably changing twenty-first-century real world, your Ph.D. journey—like the real world—may have much to do with what is hidden. To this end, the following are hidden Ph.D. stuff worthwhile reflecting on.[1]

! Attention: Hidden Ph.D. Stuff Worth Pondering At

- *Where is the uncharted scholarly territory concerning your research field?*
- *Which one is the relatively uncharted scholarly territory where you can publish many incremental works of limited impact?*
- *Which one is the uncharted scholarly territory where you can publish a few fundamental works of significant impact?*
- *Which scholarly territory is convenient for publishing theory works?*
- *Which scholarly territory is convenient for publishing implementation works?*
- *Where are the possibilities for innovation regarding your research field?*
- *How much is your Ph.D. (co-)supervisor expecting?*
- *Where to start from to maximize your scholarly impact?*
- *How can you impact your assigned Ph.D. research topic most?*
- *Can you make an impact through applied research, basic research, or a combination of them?*
- *How exactly is the peer review system working?*
- *How to get efficiently published?*
- *How to surely convince your most challenging anonymous technical reviewer?*

Oftentimes, hidden things that we come across in life (a.k.a. *the real world*) are the ones that would make us regret—in one way or another—for failing to foresee them in advance. Had we succeeded in anticipating them through intensive thinking and imagination, we would have enjoyed the smoothest ride in life and a research

[1] As a Ph.D. student and new researcher, your job shall be all about discovering new scientific phenomena—i.e., new theories, algorithms, tradeoffs, and realization—and uncovering hidden stuff that can be *interconnected and integrated* in a novel way. To fulfill such expectations, you may often sail in uncertain, unknown, and random scholarly territories to disseminate new knowledge and innovation to your disciplinary research community and the broader scientific research community. Thus, the most important question for a new researcher and emerging thinker is: *How can new knowledge and innovation that would have the most impact be efficiently and frequently delivered?*

career. To this end, being *a critical thinker, a creative thinker, a strategic thinker, a probabilistic thinker, a causal thinker, and an independent thinker* is the ultimate solution that would serve any unpredictable journey—like your intellectual journey in a Ph.D. study—very well. *Critical, creative, strategic, probabilistic, causal, and independent thinking* are the most important skills any researcher should develop routinely. They are also the most important intellectual qualities that can enable you to accomplish anything, let alone your Ph.D. degree. On the contrary, they are also the ones you have never been taught of—with a high probability—in your undergraduate or graduate courses. They also necessitate your persistent and intentional endeavor to develop as a critical thinker, a creative thinker, a strategic thinker, a probabilistic thinker, a causal thinker, and an independent thinker.

Unfortunately, developing as a critical thinker, a creative thinker, a strategic thinker, a probabilistic thinker, a causal thinker, and an independent thinker forces anyone to go through routine developmental processes, including "failures."[2] This would thus lead to many *irreducible chicken-and-egg problems* that give rise to generic Ph.D. pitfalls, which are critical by nature. Specifically, generic Ph.D. pitfalls are Ph.D. pitfalls that can happen throughout one's Ph.D. study, such as maintaining a work-life balance, intensive engagement in life, your expectations, your Ph.D. (co-)supervisor's expectations, maintaining your curiosity, a lack of comprehensive research skills, and developing as a critical thinker, a creative thinker, a strategic thinker, a probabilistic thinker, a causal thinker, and an independent thinker, which are all inevitable throughout your Ph.D. study.

As to the inevitability of various generic Ph.D. pitfalls, this chapter offers considerable attention to how and why they occur for the following reasons: first, to inform you in advance of the likely generic Ph.D. pitfalls that you would face in your Ph.D. study; second, to prepare you in advance so that you will not get bogged down by the inevitability of various generic Ph.D. pitfalls throughout your Ph.D. study; third, to motivate you to turn proactively to the references cited by Chap. 7 and this book; fourth, to encourage you to discuss the generic Ph.D. pitfalls with your network of peers; and, last, to inspire you to develop critical thinking, creative thinking, strategic thinking, probabilistic thinking, causal thinking, and independent thinking capabilities, which you would need throughout your Ph.D. study and beyond.

Following this introduction, the remainder of this chapter is organized as follows: Sects. 7.2–7.7 discuss various generic pitfalls of a Ph.D., such as maintaining a work-life balance, intensive engagement in life, your expectations, your Ph.D. (co-) supervisor's expectations, maintaining your curiosity, and a lack of comprehensive research skills, respectively; Sect. 7.8 details the generic pitfalls of a Ph.D. with respect to (w.r.t.) a lack of experience in original thinking, which translates to the generic Ph.D. pitfalls of *developing as a critical, a creative, a strategic, a probabilistic, a causal, and an independent thinker*; and, last, Sect. 7.9 provides concluding remarks.

---

[2] Several thinkers and researchers agree that there are no *"failures," but wrong results*!

7.2 Maintaining a Work-Life Balance

> *Things which matter most must never be at the mercy of things which matter least.*
>
> —*Goethe (1749–1832)*

For any Ph.D. student and new researcher such as yourself, satisfying the high intellectual standards of research, writing, and publication translates to persistent hard work that should be executed by a motivated mind with a purpose in science and innovation. Apart from persistent hard work, you need to overcome several inevitable Ph.D. pitfalls, such as research-related, writing and publication-related, management-related, persona-related, and motivation-related pitfalls. By surmounting these pitfalls, the broader scientific community—as your prospective anonymous reviewers—would expect you to disseminate high-impact research in top-notch journals while committing yourself to spend your time generously on scientific research. As a result, you would likely have difficulty maintaining a work-life balance throughout your entire Ph.D. study, giving rise to the entitled genetic Ph.D. pitfall.

Throughout your Ph.D. study, you may fail to maintain a work-life balance for several reasons. First, you can be burdened by the need to conduct several research investigations to succeed in publishing quality journal papers, forcing you to either overlook or ignore aspects of your work-life balance. Second, you may lack depth and breadth of expertise in your research field, forcing you to spend much of your time building expertise in your research field while overlooking or ignoring aspects of your work-life balance. Third, the high expectations of your Ph.D. (co-)supervisor and graduate school may force you to work non-stop while absorbed, leaving you no time for a work-life balance. Fourth, you may not have a foundational work to build on, forcing you to make substantial endeavors throughout your Ph.D. study while spending much of your time. Fifth, you may have limited or no advanced professional writing and publication experience, forcing you to work hard while causing a work-life imbalance. Sixth, you may sometimes be forced to work overnights and weekends due to complex research problems to solve, multiple deadlines, and tough journal paper revisions. Seventh, the exceptionally high intellectual standards of many top-tier journals would implicitly urge you to work numerous hours every week to the extent that you would be unable to balance your work and life. Last, you can be stressed by the never-ending demand for hard work and winning the intense competition in your research field, so you would either overlook or ignore many aspects of work-life balance. Accordingly, the entitled generic Ph.D. pitfall is inevitable throughout your Ph.D. study.

To be a successful Ph.D. student and new researcher, you may need to work creatively and effectively very hard. However, as a Ph.D. student and new researcher, you may not have the skills and expertise to work creatively and effectively hard so that you would be able to create time for your work-life balance. Therefore, the entitled generic Ph.D. pitfall boils down to numerous irreducible *chicken-and-egg problems*, which can trouble you throughout your entire Ph.D. study.

We now proceed with a generic Ph.D. pitfall of intensive engagement in life.

7.3 Intensive Engagement in Life

With self-discipline, most anything is possible.

—Theodore Roosevelt (1858–1919)

As discussed in the previous chapters, being a Ph.D. student entails several commitments and constraints, such as persistent hard work, a busy professional life, a tightly scheduled personal life, proactivity, adaptivity, resiliency, and so forth, which shape your *introverted life*—as a new researcher—throughout your Ph.D. study. On the other hand, intensive engagement in your personal life constitutes the extroverted life that can bring you several friends, abundant connections, relaxation, memorable youth times, and the like. Unfortunately, nevertheless, these seemingly interesting rewards often come at the cost of being engaged in tasks and actions that are either *not important and urgent* or *not important and not urgent*[3] while considering that your aims as a Ph.D. student are all about disseminating high-impact journal papers and earning your Ph.D. degree. Your probability of spending time on such tasks and actions would be considerably high as you get more and more engaged in your extroverted life while implicitly making you evade the essential Ph.D. tasks and actions that are either *important and urgent* or *important and not urgent*. This is how and why the entitled generic Ph.D. pitfall can happen during your entire Ph.D. study, spelling a massive disaster for any international Ph.D. student who may not know the abroad community and environment one will be co-existing with.

Aiming to adapt to a new environment and community as a local or international Ph.D. student, you may engage with your community living in a city/town you would be researching. Doing so may help you in managing emotion and stress, gaining new life experience, getting helpful local information, and alike, though these pros come at the cost of wasting valuable time via too much engagement, getting distracted by too much engagement, sailing in a wrong environment that can be a viable cause for demotivation, etc. You may engage electronically with your members back home to manage emotion and stress, maintain family connection and participation, receive family love and care, and so on, though these pros come with the cons of motivating too much engagement that can regularly waste your

[3] As to the urgency and importance of your tasks and actions, they can be classified into four quadrants (*Q*s): *QI, important and urgent tasks & actions*; *QII, important and not urgent tasks & actions*; *QIII, not important and urgent tasks & actions*; and *QIV, not important and not urgent tasks & actions* [1–3].

valuable research time and cause nostalgia, which further impacts your constrained research time and attention. Generally, you may engage with your community and family because of the following incentives for *being available in the world of others*: facilitating a work-life balance and a means of establishing a support system. Such incentives can knowingly or unknowingly lure you to intensively engage in life throughout your Ph.D. study, ascertaining the inevitability of the entitled generic Ph.D. pitfall during your entire Ph.D. study.

Since the (distracted) world of others—whether they are family members, friends, relatives, or community members who are not researchers—are *quite different* from an introverted and quiet research world, typical of Ph.D. students, you shall be mindful that you cannot succeed by being in both worlds equally. Even if you may try, you cannot succeed—with a high likelihood—as much as a research student who concentrates on one's own quiet and introverted research world. This is why you may need to avoid (mindless) social media for most of your Ph.D. study per the suggestion of Cal Newport, who is the author of a book entitled *Deep Work: Rules for Focused Success in a Distracted World* [4]. Meanwhile, to be successful with your Ph.D. study, you need to engage more with your introverted life while engaging much less with your extroverted life. Nonetheless, as a Ph.D. student and new researcher, you may not have adequate experience in effectively engaging more with your introverted life and much less with your extroverted life. Therefore, the entitled generic Ph.D. pitfall comes down to numerous irreducible chicken-and-egg problems, which can trouble you throughout your Ph.D. study.

We now continue with the next generic Ph.D. pitfall of your expectations.

7.4 Your Expectations

One of the most important results you can bring into the world is the you that you really want to be.

—Robert Fritz (1943–)

You would be pretty motivated to conduct your Ph.D. research whenever you have high expectations since it is one of the significant factors that can spark your inherent motivation [5].[4] However, too much expectation can cause immense frustration when things are not turning out in your favor, which is often the case in a Ph.D. study.[5] When you have low expectations, on the other hand, you may not be

[4] Note that [5] is a self-help book whose author had conducted decades of research on it. Although the academic community may not respect self-help books because of their insufficient evidence, we refer to [5] because we consider the book a useful resource featuring valuable advice and declarations.

[5] Because a Ph.D. journey is an unpredictable intellectual expedition full of pitfalls, witnessing and living the unexpected can be common in anyone's Ph.D. journey.

sufficiently motivated—or even demotivated—to pursue your Ph.D. study, so any little "failure" such as a paper rejection can be severely depressive and unbearable, minimizing your likelihood of success by limiting your scholarly adventure in your Ph.D. research. Hence, the entitled is a generic Ph.D. pitfall that can happen throughout your Ph.D. study depending on the consciousness manifested by your mind triggered by different circumstances.

Since conducting research is also an emotional experience, your expectancy for your Ph.D. research can swing as much as your emotions throughout your Ph.D. study. On the one hand, your mind may manifest high expectancy for your Ph.D. research when you feel motivated, empowered, energized, tireless, highly interested in your research, adventurous in your research, in control, generating new ideas, accelerating, enduring, and more introverted and focused. On the other hand, your mind may exhibit low expectancy for your Ph.D. research when you feel demotivated, depressed, exhausted, lazy, helpless, uninterested in your research, less adventurous in your research, having no new ideas, complaining, procrastinating, more extrovert and distracted, and hanging out frequently with your friends, which can all lead to stress and unhappiness throughout your Ph.D. study. Accordingly, the entitled generic Ph.D. pitfall is inevitable throughout your Ph.D. study.

While *ambition and expectancy* can spark your motivation [5], there shall always be a delicate balance between idealism and realism since your Ph.D. journey will not be straightforward or predictable. To this end, both too much or too little expectancy is surely a generic Ph.D. pitfall[6] that can be counterproductive for your research progress throughout your Ph.D. study. Meanwhile, to be a very successful Ph.D. graduate, you may need to expect a lot from yourself so that you can also spark your inherent motivation throughout your Ph.D. study. As a Ph.D. student and new researcher, however, expecting a lot from yourself can cause significant anxiety and stress. Therefore, the entitled generic Ph.D. pitfall comes down to many irreducible chicken-and-egg problems, which you may face throughout your Ph.D. study.

We now move on to another generic Ph.D. pitfall of your Ph.D. (co-)supervisor's expectations.

7.5 Your Ph.D. (Co-)Supervisor's Expectations

Expect the best, plan for the worst, and prepare to be surprised.

—Denis Waitley (1933–)

Since a Ph.D. degree is awarded for mastery of your research subject, demonstration of independent research and insight, and ability to communicate your results while relating them to a broader treatise [6], almost all supervising professors have

[6] Optimized expectations, such as result-driven expectations, can be a key to overcoming this pitfall.

high expectations—by default—for the research of their Ph.D. advisees, enforcing persistent and creative hard work on Ph.D. students. Because you may not be fully ready for or fully aware of such substantial endeavors that must be continually demonstrated, your Ph.D. (co-) supervisor's high expectations can constantly be daunting. Consequently, what implicitly becomes your troubling Ph.D. pitfall is your Ph.D. (co-)supervisor's high expectations, which are opportunities for conducting high-impact research. This is how and why the entitled generic Ph.D. pitfall occurs throughout your Ph.D. study.

You may not be fully ready to demonstrate significant professional commitments commensurate with the high expectations of your Ph.D. (co-)supervisor for various reasons. First, you may not know precisely the significant amount of persistent effort needed to catch up with your Ph.D. (co-)supervisor's high publication standards. Second, you may not know the unrelenting efforts required to rise to your Ph.D. (co-)supervisor's high scholarly caliber. Third, the high expectations of your Ph.D. (co-)supervisor and the inherent uncertainty of a Ph.D. study can stress you too much. Fourth, you may lack substantial research experience to rise to the challenges of your Ph.D. (co-)supervisor's high expectations. Fifth, you may not have considerable experience in creativity and innovation to qualify for the high expectations of your Ph.D. (co-)supervisor. Last, you have little to no publication experience to satisfy your Ph.D. (co-)supervisor's high publication standards. Accordingly, the entitled generic Ph.D. pitfall is unavoidable throughout your Ph.D. study.

Because your expert Ph.D. (co-)supervisor is a curious researcher whose most professional life has been spent on conducting original research, winning competitive grants, writing high-quality (also award-winning) journal and conference papers, supervising and directing innovative research, peer-reviewing, editing, and leading in your research field, such an ultra-high qualification needs to be reciprocated with substantial endeavors and commitments, which you should invest significant time and effort for.[7] Meanwhile, to be successful in your Ph.D. study, your Ph.D. research needs to be driven and challenged by the high expectations of your Ph.D. (co-)supervisor. Nevertheless, as a Ph.D. student and new researcher, you may not always have sufficient research experience and capability to rise to the challenges of your Ph.D. (co-)supervisor's high expectations. Therefore, the entitled generic Ph.D. pitfall comes down to various irreducible chicken-and-egg problems, which you can face throughout your Ph.D. study.

We now continue with the generic Ph.D. pitfall of maintaining your curiosity.

[7] Generally, counteracting high expectations with substantial unrelenting endeavors—inherent in your high self-expectancy—is a way to surmount any generic Ph.D. pitfall that can arise from your Ph.D. (co-)supervisor's high expectations.

7.6 Maintaining Your Curiosity

Curiosity demands that we ask questions that we try to put things together and try to understand this multitude of aspects as perhaps resulting from the action of relatively small number of elemental things and forces acting in an infinite variety of combinations.

—*Richard P. Feynman (1918–1988)*

As for the goals of many research projects various Ph.D. students would be working on and the expectation of the respective supervising professors, there is a general contemporary academic trend that Ph.D. students are expected to be result-oriented research assistants who target papers most of the time. Contrary to such a common trend in current academia, history reminds us that the most groundbreaking scientific discoveries—such as *quantum mechanics and quantum electrodynamics*—are the results of mere *curiosity-driven research*. Given curiosity-driven research and the realities of result-driven contemporaneous academia, a discrepancy often arises in the mind of a curious Ph.D. student as to whether one should serve one's curiosity or the process of delivering results for the sake of delivering results. While serving one's curiosity shall always be the primary purpose of any researcher, the limited-time funding reality of your Ph.D. study—i.e., *the reality that bites*—can force you to breathlessly stick to the routine of delivering results, making any attempt to maintain your curiosity nearly impossible, if not completely impossible. This is how and why the entitled generic Ph.D. pitfall surely happens throughout your Ph.D. study.

Curiosity and sustained curiosity during graduate study are the major factors that draw several graduate students into Ph.D. programs. Despite the different intrinsic expectations of curious minds, the contemporary result-driven research enterprise may forcefully drag them into a different culture and expectation to the extent that they would be uninterested in research and what they were curious about. Such disillusionment may force Ph.D. students to be only result-oriented researchers while compelling them to give up on their curiosities. However, giving up on one's curiosity amounts to moving away from a *path of novelty* that can lead to groundbreaking discoveries and impactful innovations. To this end, should research impact and the number of curious new-generation researchers be the main scientific progress metrics, the contemporary incremental and result-driven academia may have already self-defeated its purpose. In fact, what is fundamentally changing is the present-day research enterprise and academia, as also echoed by two renowned scientists and brothers who authored an entertaining PNAS opinion article entitled *Science in the age of selfies* [7]. According to the experiential wisdom of these distinguished scholars and leaders, the twenty first-century academia has mainly become a "small idea factory," where authors search for "minimum publishable units" while being continually rewarded for publishing more repeatedly [7, p. 9386]. On the contrary, according to these leading experts, this is not precisely what a curious young researcher is supposed to be doing [7].

If one has to stick to *realism rather than idealism*, what is mainly expected of a new researcher in present-day academia is tracking and contributing to your research field's state-of-the-art developments and trends. As a result, you may struggle to maintain your curiosity throughout your Ph.D. study since contemporary academia would directly or indirectly steer your Ph.D. research project. You may also struggle to maintain your curiosity throughout your Ph.D. study since following your curiosity can increase your risk of being quickly outdated concerning the ever-evolving state-of-the-art advancements of any twenty first-century research discipline. You may also struggle to maintain your curiosity throughout your Ph.D. study because following your curiosity can introduce considerable uncertainty to the goals of your Ph.D. project that must be achieved in a few years of funded research. You may also struggle to maintain your curiosity throughout your Ph.D. study since you could have no extra time left while under immense pressure to deliver original results to your funded Ph.D. research project. You may also struggle to maintain your curiosity throughout your Ph.D. study because your curiosity-driven exploration may not be funded by a funding agency that sponsors your Ph.D. (co-)supervisor's research project. You may also struggle to maintain your curiosity throughout your Ph.D. study since you may not be fortunate enough to have a Ph.D. (co-)supervisor who encourages and supports your immense research freedom. You may also struggle to maintain your curiosity throughout your Ph.D. study because you do not have independent funding to follow only your curiosity. Accordingly, the entitled generic Ph.D. pitfall is inevitable throughout your Ph.D. study.

Summing up, being too result oriented or pushed for the sake of papers can force you to give up on your intrinsic curiosity. Hence, there shall be a *sweet spot* between being curiosity driven and being result oriented. Otherwise, the entitled generic Ph.D. pitfall can harm your long-term career advancement prospects, though it may have no impact in the short term. Meanwhile, to be remarkably successful in your Ph.D. study, you need to have unbounded curiosity throughout your Ph.D. study. Nonetheless, as a Ph.D. student and new researcher, you may not have adequate funding, sufficient time, and complete independence to maintain your unbounded curiosity. Therefore, the entitled generic Ph.D. pitfall boils down to many irreducible chicken-and-egg problems, which can confront you during your entire Ph.D. study.

We now proceed with the generic Ph.D. pitfall of a lack of comprehensive research skills.

7.7 A Lack of Comprehensive Research Skills

You have to have absolute confidence. Keep right on going, and nothing will happen.

—Richard P. Feynman (1918–1988)

If a new Ph.D. student has significant prior research experience and expertise in one's Ph.D. research, getting a Ph.D. can get easier to some extent. In any Ph.D.

study that deals with problem-based research, arriving at a published journal paper entails many consecutive activities and tasks that should be executed by a motivated mind with significant clarity, precision, criticality, and creativity. The major ones are rigorous and intuitive understanding of your research field's state-of-the-art research advancements; imagination and generation of novel ideas; formulation of a novel and feasible research problem; devising novel solutions—with regard to a suitable mathematical framework—to the formulated problem; validation of the devised solutions using numerical simulations or computer experiments; comparing the devised solution with the existing solutions, if any; writing, re-writing, and editing a journal paper; submission of the written and edited journal paper; revising and editing the submitted journal paper to address the anonymous reviewers' comments; and finally, getting it published. Succeeding in all these intellectual activities and tasks undoubtedly requires comprehensive research skills, which you naturally and unfortunately lack as a Ph.D. student and new researcher who may also lack significant prior research experience.[8] This is why the entitled generic Ph.D. pitfall arises throughout your Ph.D. study.

According to the book in [8], to perform a good research in the natural sciences (also engineering), a researcher must draw on an inquisitive mind, an appreciation of the aims and methods of science (and engineering), the limitations of science (and engineering), and the skill in exploiting the "tools of the trade" for science and engineering [8, p. v]. However, for many reasons, you may consistently face challenges due to a lack of comprehensive research skills throughout your Ph.D. study. First, your lack of comprehensive research skills can be compounded by your lack of subject-matter expertise, making your Ph.D. journey typically challenging. Second, you may lack time to *adaptively, proactively, and curiously* learn new knowledge on the rapid state-of-the-art developments of your research field and its limitations. Third, you may lack adequate time and sufficient motivation to learn old and new methods, which are constantly being developed. Fourth, you may need abundant time to master the "tools of the trade," since the number of such tools continues to rise. Fifth, the underlying *time-varying randomness* (a.k.a. *indeterminism*) in your research field's knowledge, tools, methods, and limitations. Sixth, you can be under immense time pressure only to execute result-driven research to be regularly reported to funding agencies rather than spending ample time to develop comprehensive research skills. Last, you may lack considerable scholarly experience in research, writing, publication, management, and original thinking. Consequently, the entitled generic Ph.D. pitfall is inevitable throughout your Ph.D. study.

Acquiring comprehensive research skills requires proactive and persistent engagement *in depth and breadth* of research within the scope of your Ph.D. research field and beyond. Besides, it necessitates—among others—continuous mind development as a critical thinker, a creative thinker, a strategic thinker, a probabilistic thinker, a causal thinker, and, most importantly, an independent

---

[8] We shall underscore that one would get experience after one needs it.

thinker. Meanwhile, you must have comprehensive research skills to have a smooth Ph.D. journey and succeed in your Ph.D. study. However, as a new Ph.D. student and researcher, you may lack comprehensive research skills. Therefore, the entitled generic Ph.D. pitfall comes down to several chicken-and-egg problems, which can trouble you throughout your Ph.D. study.

We now move on to the generic Ph.D. pitfall of developing as an original thinker.

7.8 Developing as an Original Thinker

Thinking is the hardest work there is, which is probably the reason why so few engage in it.

—Henry Ford (1863–1947)

Being a Ph.D. student and new researcher makes you confronted by several research-related Ph.D. pitfalls, writing-related Ph.D. pitfalls, publication-related Ph.D. pitfalls, management-related Ph.D. pitfalls, persona-related Ph.D. pitfalls, and motivation-related Ph.D. pitfalls, which are all inevitable during the different phases of your Ph.D. study. While dealing with all these Ph.D. pitfalls that can also happen concurrently, you need to effectively manage the various scientific and non-scientific issues of your Ph.D. study, such as psychology management, time management, stress management, happiness and health management, motivation management, research management, focus management, environment (sociological) management, productivity management, etc.

By overcoming the many diverse Ph.D. pitfalls and effectively managing the various scientific and non-scientific issues of your Ph.D. study, you are usually expected to publish several high-quality journal and conference papers, which would be the foundation for your prospective Ph.D. dissertation. To succeed at this smoothly and elegantly, you need to quickly develop as an original thinker endowed with the capability of original thinking. Original thinking routinely pursued from the viewpoint of criticality is known as critical thinking; from the viewpoint of creativity, it is called creative thinking; from the viewpoint of strategy, it is termed strategic thinking; from the viewpoint of probability, it is named as probabilistic thinking; from the viewpoint of causality, it is called causal thinking; and from the viewpoint of independence and autonomy, it is termed independent thinking. Because one's routine engagement in thinking independently and autonomously can hardly succeed without criticality, creativity, strategy, probability, and causality, independent thinking entails critical thinking, creative thinking, strategic thinking, probabilistic thinking, and causal thinking.

Amid your endeavor to develop as an original thinker, you would face various original thinking-related generic Ph.D. pitfalls throughout your Ph.D. study as a Ph.D. student and new researcher for several reasons. First, your prior education and training may not have offered you ample intellectual opportunities that lay the foundation for your development as a critical thinker, a creative thinker, a strategic thinker, a probabilistic thinker, a causal thinker, and an independent thinker. Second, you may lack sufficient intellectual experience in original thinking. Third, you may lack depth and breadth of subject-matter expertise and perspectives that can inform and steer original thinking. Fourth, you may lack significant experience in problem-solving using original thinking. Last, you may have a hectic Ph.D. life ridden with deadlines that commonly cause stress, detrimentally impacting the quality of your developing original thinking. Consequently, the following generic Ph.D. pitfalls are inevitable throughout your Ph.D. study.

! Attention: Generic Pitfalls w.r.t. a Lack of Experience in Original Thinking

- *Developing as a critical thinker.*
- *Developing as a creative thinker.*
- *Developing as a strategic thinker.*
- *Developing as a probabilistic thinker.*
- *Developing as a causal thinker.*
- *Developing as an independent thinker.*

In what follows, we detail each of the itemized generic pitfalls of a Ph.D.

7.8.1 Developing as a Critical Thinker

If two men agree on everything, you may be sure that one of them is doing the thinking.

—Lyndon B. Johnson (1908–1973)

Being a Ph.D. student and new researcher naturally makes you face many diverse Ph.D. pitfalls during your entire Ph.D. study. What is exceedingly challenging about the Ph.D. pitfalls is that they can happen simultaneously to the extent that applying specific overcoming strategies for each can hardly be an effective overall strategy. However, you can devise thoughtful surmounting strategies adaptively and persistently, provided that your mind develops as a critical thinker. What also entails intellectual development as a critical thinker is publishing several journal and conference papers as a first author since you have to think critically to handle different anonymous technical reviewers properly. Managing the various scientific and non-scientific issues of your Ph.D. study also entails intellectual development as a critical thinker. Accordingly, to successfully overcome the intellectual challenges of your Ph.D. study, you would undoubtedly need your development as a critical thinker. Nonetheless, successfully developing as a critical thinker can be intellectually challenging for any Ph.D. student and new researcher, underlining the entitled generic Ph.D. pitfall.

To any Ph.D. student and new researcher such as yourself, developing as a critical thinker can be challenging because your prior education and training may not have offered you abundant intellectual opportunities that can pave the way to your development as a critical thinker. It can also be challenging since you may lack significant intellectual experience in solving problems—i.e., personal and research problems—using critical thinking. It can also be challenging since you may lack significant intellectual experience in critical thinking. It can also be challenging since critical thinking requires intensive thinking through various depth and breadth of issues [9–11], which you may hardly succeed at for you may lack—as a new researcher—depth and breadth of expertise in your research field. It can also be challenging since critical thinking requires intensive thinking via numerous points of view [9–11], which you may lack as a new researcher in your research field. It can also be challenging since critical thinking requires intensive thinking through astronomically many concepts [9–11], which you may lack as a new researcher in your research field. It can also be challenging since you may not have ample time to spend on your development as a critical thinker, considering your packed Ph.D. schedule filled with various responsibilities. Consequently, the entitled generic Ph.D. pitfall is inevitable throughout your Ph.D. study.

In the context of critical thinking, the *intellectual standards* are clarity, accuracy, relevance, logicalness, breadth, precision, significance, completeness, fairness, and depth; the *elements of reasoning* are purposes, questions, points of view, information, inferences, concepts, implications, and assumptions; and the *intellectual traits* are intellectual humility, intellectual autonomy, intellectual integrity, intellectual courage, intellectual perseverance, confidence in reason, intellectual empathy, and fairmindedness [9–13]. To this end, to develop as a critical thinker, you may need to have sufficient intellectual experience in applying the intellectual standards repeatedly to the elements of reasoning to acquire the intellectual traits [9–13]. As a Ph.D. student and new researcher, however, you may lack sufficient intellectual experience and subject-matter expertise to successfully apply intellectual standards to the elements of reasoning, making the development of intellectual traits difficult. Therefore, the entitled generic Ph.D. pitfall boils down to several irreducible chicken-and-egg problems, which can trouble you throughout your Ph.D. study.

7.8.2 Developing as a Creative Thinker

To my mind, it is a mistake to think of creative activity as something unusual.

—Jacob Bronowski (1908–1974)

Surmounting the many diverse Ph.D. pitfalls—that you would face simultaneously throughout your Ph.D. study—entails the capability of not only critical thinking but also creative thinking. Apart from critical thinking, you also need creative thinking to manage the many scientific and non-scientific issues of your Ph.D.

study and successfully publish various journal and conference papers. While there is no substitute for hard work, novel results can often happen due to the germination of novel ideas. Accordingly, each step of your Ph.D. research and daily scientific activities would likely be successful provided that you conduct them with considerable creativity, as a manifestation of novel creative thinking. Nevertheless, successfully developing as a creative thinker can be intellectually challenging for any Ph.D. student and new researcher such as yourself, calling forth the entitled generic Ph.D. pitfall.

To any Ph.D. student and new researcher such as yourself, developing as a creative thinker can be challenging since your prior education and training may not have offered you sufficient intellectual opportunities to lay the foundation for your development as a creative thinker. It can also be challenging since you may lack significant intellectual experience in creative thinking; since creative thinking can require *non-conformist thinking* [14], in which you may lack adequate intellectual experience; since creative thinking can require various cycles of *divergent and convergent thinking* [15], in which you may not have ample intellectual experience; since you may lack significant scholarly experience in generating ideas, which can be a stepping stone for your *creative thinking* [16]; since you may lack adequate intellectual experience in imagination and imaginative thinking that pave the way to considerable creativity and creative thinking; since your huge workload and its ensuing stress throughout your Ph.D. study can diminish your creativity because stress can impact anybody's cognitive abilities and creativity; and since you may not have ample time to spend on your development as a creative thinker, considering your busy Ph.D. schedule filled with many responsibilities. As a result, the entitled generic Ph.D. pitfall is unavoidable throughout your Ph.D. study.

Your creative thinking would develop provided that you engage yourself in originating several creative works. Nevertheless, such creative works often require intensive creative thinking to be creative enough. Meanwhile, novel ideas that usually motivate creative works can be generated by your mind, provided that you manage to develop yourself as a creative thinker through a gigantic intellectual experience of creativity. However, as a Ph.D. student and new researcher, you may lack the intellectual experience of creativity to develop yourself successfully as a creative thinker. Therefore, the entitled generic Ph.D. pitfall comes down to many irreducible chicken-and-egg problems, which you may face throughout your Ph.D. study.

7.8.3 Developing as a Strategic Thinker

A vision without a strategy remains an illusion.

—Lee G. Bolman (1941–)

Throughout your Ph.D. study, almost everything requires a thoughtful set of strategies. For instance, you would need thoughtful literature reading/tracking strategies, self-motivation strategies, publication strategies, innovation strategies, work-life balance strategies, high-effectiveness/high-performance strategies, curiosity-maintenance strategies, collaboration strategies, and paper writing strategies. Hence, you certainly need many adaptive strategies to be applied per the indeterministic situation at hand because there is no *one-size-fits-all strategy*, calling for your intensive strategic thinking. In this vein, overcoming the many diverse pitfalls entails a capability of not only critical thinking and creative thinking but also strategic thinking. Aside from critical thinking and creative thinking, you also need strategic thinking to overcome the several Ph.D. pitfalls that can co-occur; manage the many scientific and non-scientific issues of your Ph.D. study; and successfully publish various journal and conference papers. On the contrary, you may struggle to develop as a strategic thinker because thoughtful strategic thinking can be intellectually challenging for a Ph.D. student and new researcher, affirming the entitled generic Ph.D. pitfall.

Strategy is a means to an end [17]; in your case, it is a means to your Ph.D. degree. To any Ph.D. student and new researcher such as yourself, developing as a strategic thinker can be challenging since your prior education and training may not have offered you plenty of intellectual opportunities that can set the stage for your development as a strategic thinker. It can also be challenging because you may lack significant intellectual experience in thoughtful strategic thinking; because you may lack scholarly experience in critical and creative thinking that can facilitate your development as a strategic thinker; since thoughtful strategic thinking also requires thinking through the vantage point of *a strategy for a strategy* [17]; because you may lack considerable pragmatic experience, knowledge, and wisdom to think through the lens of strategy; because you may lack ample strategic wisdom to think deeply using strategy as an overarching lens; and since you may not have sufficient time to spend on your development as a strategic thinker, considering your busy Ph.D. schedule filled with numerous responsibilities. Accordingly, the entitled generic Ph.D. pitfall is inevitable throughout your Ph.D. study.

As a Ph.D. student who is expected to deliver an original Ph.D. dissertation in 3–5 years, you naturally have limited time, energy, knowledge, funding, and wisdom.[9] Your success in your Ph.D. hinges on the effective management and optimization of these constraints. Such management and optimization call for many adaptive strategies informed and steered by significant research experience. However, as a Ph.D. student and new researcher, you may not have substantial research experience to efficiently optimize the limited resources mentioned while endeavoring to get things done optimally. Therefore, the entitled generic Ph.D. pitfall boils down to many irreducible chicken-and-egg problems, which can trouble you throughout your Ph.D. study.

[9] Wisdom is applied knowledge; see the *data, information, knowledge, and wisdom (DIKW) pyramid* [18, p. 55, Figure 2].

7.8.4 Developing as a Probabilistic Thinker

Life is a school of probability.
—Walter Bagehot (1826–1877)

From an *idealistic vantage point*, a given principle or strategy is right. However, from a *realistic vantage point*, it can lead only to a tiny probability of success. Should you thus aim to maximize your probability of success in your Ph.D. study, you need not only thoughtful strategies but also intensive engagement in probabilistic thinking, which is all about thinking through the fascinating lens of probability. In this regard, surmounting the many inevitable Ph.D. pitfalls entails a capability of not only critical thinking, creative thinking, and strategic thinking but also probabilistic thinking. Apart from critical thinking, creative thinking, and strategic thinking, you also need probabilistic thinking to overcome the several Ph.D. pitfalls that can occur simultaneously; manage the many scientific and non-scientific issues of your Ph.D. study; and successfully publish various journal and conference papers. Nevertheless, you may toil to develop as a probabilistic thinker since contemplative probabilistic thinking can be intellectually challenging for any Ph.D. student and new researcher such as yourself, asserting the entitled generic Ph.D. pitfall.

Probabilistic thinking is aimed at accurately estimating a chance of success strategized by idealistic and realistic principles. To any Ph.D. student and new researcher such as yourself, developing as a probabilistic thinker can be challenging for many reasons. First, your prior education and training may not have offered you abundant intellectual opportunities that can prepare the ground for your development as a probabilistic thinker. Second, you may lack considerable intellectual experience in probabilistic thinking. Third, you may lack significant scholarly experience in critical, creative, and strategic thinking that can facilitate your development as a probabilistic thinker. Fourth, you may lack considerable pragmatic experience, knowledge, and wisdom to think through the lens of probability. Fifth, you may lack adequate probabilistic wisdom to think intensively using probability as a primary lens. Last, you may not have ample time to spend on your development as a probabilistic thinker, considering your busy Ph.D. schedule filled with many responsibilities. Consequently, the entitled generic Ph.D. pitfall is unavoidable throughout your Ph.D. study.

Routinely engaging in probabilistic thinking usually offers you the opportunity to foresee—with some probability of accuracy—the trajectory of your Ph.D. success as being lined up with hurdles or smooth avenues, enabling you to pass the right decision at the right time regardless of a sound vehicle of strategy being chosen. Beyond probabilistic principles, developing as a probabilistic thinker entails outstanding strategic, creative, and critical thinking, which you may struggle to qualify for since you have not yet developed yourself as a creative thinker, critical thinker, and strategic thinker. On the other hand, considering any event's chance of randomness, there are infinite manifestations of any random event since there are infinite numbers between 0 (uncertain likelihood) and 1 (certain likelihood). Comprehending any random event amounts to thinking probabilistically through infinite manifestations of a random event. Nonetheless, as a Ph.D. student and new researcher, you may lack gigantic probabilistic wisdom to think through nearly countless manifestations of a random event. Therefore, the entitled generic Ph.D. pitfall comes down to many irreducible chicken-and-egg problems, which can trouble you throughout your Ph.D. study.

7.8.5 Developing as a Causal Thinker

Shallow men believe in luck or in circumstance. Strong men believe in cause and effect.

—Ralph Waldo Emerson (1803–1882)

We sometimes tend to cause things unknowingly and understand the effect of our actions only after observing their respective results that are counterproductive. The undesirable results justify abandoning those actions as much as one needs to engage in rigorous causal thinking before taking any action. Should you view the many diverse and inevitable Ph.D. pitfalls as the (in)direct result of underlying processes of causes and effects, overcoming them entails a capability of not only critical, creative, strategic, and probabilistic thinking but also causal thinking. Aside from critical thinking, creative thinking, strategic thinking, and probabilistic thinking, you also need causal thinking to overcome the various Ph.D. pitfalls that can happen concurrently; manage the many scientific and non-scientific issues of your Ph.D. study; and successfully publish various journal and conference papers. However, you may sweat to develop as a causal thinker since rigorous causal thinking can be intellectually challenging for a Ph.D. student and new researcher such as yourself, attesting to the entitled generic Ph.D. pitfall.

Generally, causal thinking is a thinking endeavor through the lens of causality; particularly, it is a thinking effort via the lens of *causal inference (i.e., causal reasoning and causal learning)* [19–22]. To any Ph.D. student and new researcher such as yourself, developing as a causal thinker can be challenging for many reasons. First, your prior education and training may not have offered you sufficient intellectual opportunities that can pave the road to your development as a causal thinker. Second, you may lack significant intellectual experience in causal thinking. Third, you may lack considerable scholarly experience in critical, creative, strategic, and probabilistic thinking that can accelerate your development as a causal thinker. Fourth, you may lack significant pragmatic experience, knowledge, and wisdom to think through the lens of causality. Fifth, you may lack adequate causal wisdom to think deeply using causality as a primary lens. Last, you may not have ample time to spend on your development as a causal thinker, considering your busy Ph.D. schedule filled with many responsibilities. Accordingly, the entitled generic Ph.D. pitfall is inevitable throughout your Ph.D. study.

In view of causal thinking, you may develop a mindset of cause and effect toward overcoming any Ph.D. pitfall you may encounter during your Ph.D. study. Specifically, you may think of your Ph.D. issues as the effect of a causal process that has caused them. Toward your Ph.D. and beyond, you shall always attempt to answer the following question: What should I do today (*your action as a cause*) to achieve an outstanding academic, industrial, and/or entrepreneurial career (*career as a direct effect of your cumulative actions*)? Meanwhile, to develop as a causal thinker, you need pragmatic experience, knowledge, and wisdom to think through the lens of causality. As a Ph.D. student and new researcher, nevertheless, you may lack pragmatic experience, knowledge, and wisdom that can enable you to think deeply through the lens of causality and substantial intellectual experience in critical thinking, creative thinking, strategic thinking, and probabilistic thinking that can facilitate your intellectual development as a causal thinker. Therefore, the entitled generic Ph.D. pitfall boils down to various irreducible chicken-and-egg problems, which you may face throughout your Ph.D. study.

7.8.6 Developing as an Independent Thinker

What we think, we become.

—Buddha (563 BC–483 BC)

Each of the discussed thinking frameworks—i.e., critical thinking, creative thinking, strategic thinking, probabilistic thinking, and causal thinking—is invaluable to address only some aspects of your Ph.D. study. However, during your doctoral study, you may confront various circumstances requiring you to rely only on your thinking to foresee all the possibilities, perspectives, and directions. This is the essence of independent thinking you would need throughout your Ph.D. study for several purposes. You need independent thinking to surmount the many diverse Ph.D. pitfalls you will encounter throughout your Ph.D. study. You also need independent thinking to surmount the several Ph.D. pitfalls that can occur simultaneously; manage the many scientific and non-scientific issues of your Ph.D. study; and successfully publish various journal and conference papers. Nonetheless, you may struggle to develop as an independent thinker because rigorous independent thinking can be exceedingly challenging—if not impossible—for a Ph.D. student and new researcher such as yourself, underscoring the entitled generic Ph.D. pitfall.

An independent thinker is a critical thinker, creative thinker, strategic thinker, probabilistic thinker, and causal thinker who usually manifests—regardless of intellectual complexity, frustration, or strong negative emotions—substantial intellectual humility, intellectual autonomy, intellectual integrity, intellectual courage, intellectual perseverance, confidence in reason, intellectual empathy, and fairmindedness. To any Ph.D. student and new researcher such as yourself, developing as an independent thinker can be challenging because your prior education and training may not have offered you abundant intellectual opportunities that can pave the way to your development as an independent thinker. It can also be challenging since you may lack substantial intellectual experience in independent thinking; since you may lack considerable scholarly experience in critical, creative, strategic, probabilistic, and causal thinking that facilitates your development as an independent thinker; since you may lack significant pragmatic experience, depth and breadth of expertise, and sage wisdom to think independently and intensively; since you may lack adequate strategic, probabilistic, and causal wisdom to think independently and intensively; since you may lack adequate intellectual maturity to handle any intellectual complexity, frustration, and strong negative emotions; and since you may not have ample time to spend on your development as an independent thinker, concerning your packed Ph.D. schedule filled with many responsibilities. As a result, the entitled generic Ph.D. pitfall is unavoidable throughout your Ph.D. study.

Because you would be required to depend only on your thinking to foresee all the possibilities, perspectives, and directions, developing as an independent thinker requires the capability of critical thinking, creative thinking, strategic thinking, probabilistic thinking, and causal thinking. Thus, to develop as an independent thinker, you must first develop as a critical thinker, a creative thinker, a strategic thinker, a probabilistic thinker, and a causal thinker. However, as a Ph.D. student and new researcher, you have not yet successfully developed as a critical thinker, a creative thinker, a strategic thinker, a probabilistic thinker, and a causal thinker. To develop as an independent thinker, you also need strategic, probabilistic, and causal wisdom, which you may lack as a Ph.D. student and new researcher. To develop as an independent thinker, you also need considerable intellectual maturity coupled with developed expertise in both depth and breadth, which you likely lack as a Ph.D. student and new researcher. Therefore, the entitled generic Ph.D. pitfall comes down to many irreducible chicken-and-egg problems, which can confront you throughout your Ph.D. study.

7.9 Concluding Remarks

> *We cannot solve our problems with the same thinking we used when we created them.*
>
> *—Albert Einstein (1879–1955)*

This chapter detailed several generic Ph.D. pitfalls unavoidable during the initial, middle, and final years of your Ph.D. study. Specifically, we discussed various generic pitfalls of a Ph.D., such as maintaining a work-life balance, intensive engagement in life, your expectations, your Ph.D. (co-)supervisor's expectations, maintaining your curiosity, and a lack of comprehensive research skills. We also discussed the generic pitfalls of a Ph.D. w.r.t. a lack of intellectual experience in original thinking that translates to the following generic Ph.D. pitfalls: Developing as a critical thinker, developing as a creative thinker, developing as a strategic thinker, developing as a probabilistic thinker, developing as a causal thinker, and developing as an independent thinker.

Although we tried to detail the most important generic Ph.D. pitfalls inevitable throughout your entire Ph.D. study, our discussion of limited scope—due to a space constraint—may not have covered the *full-blown depth and breadth* of issues intertwined with the discussed themes. For detailed insights and realities, we encourage you to reflect this chapter's references and discuss them with your network of peers.

References

1. Covey SR, Merrill AR, Merrill RR (1994) First things first. Simon and Schuster, London
2. Covey SR (2004) The 7 habits of highly effective people: powerful lessons in personal change. Simon and Schuster, New York
3. Gabrys BJ, Langdale JA (2012) How to succeed as a scientist: from Postdoc to Professor. Cambridge University Press, Cambridge
4. Newport C (2016) Deep work: rules for focused success in a distracted world. Grand Central Publishing, New York
5. Burchard B (2014) The motivation manifesto: 9 declarations to claim your personal power. Hay House, Carlsbad
6. Petre M, Rugg G (2010) The unwritten rules of PhD research, 2nd edn. Open University Press, Berkshire
7. Geman D, Geman S (2016) Science in the age of selfies. Proc. Natl. Acad. Sci. (PNAS) 113:9384–9387
8. Pruzan P (2016) Research methodology: the aims, practices and ethics of science. Springer, Switzerland
9. Paul RW, Elder L (2002) Critical thinking: tools for taking charge of your professional and personal life. Prentice Hall, Upper Saddle River
10. Paul R, Elder L (2008) The thinker's guide for students and faculty to scientific thinking. The Foundation for Critical Thinking, Dillon Beach
11. Paul R, Niewoehner R, Elder L (2006) The thinker's guide to engineering reasoning. The Foundation for Critical Thinking, Dillon Beach
12. Paul R, Elder L (2008) The thinker's guide to the nature and functions of critical and creative thinking. The Foundation for Critical Thinking, Dillon Beach
13. Paul R, Elder L (2008) The thinker's guide to how to read a paragraph: the art of close reading. The Foundation for Critical Thinking, Dillon Beach
14. Grant A (2016) Originals: how non-conformists move the world. VIKING, New York
15. Shneiderman B (2016) The new ABCs of research: achieving breakthrough collaborations. Oxford University Press, Oxford (Kindle Edition)
16. Sternberg RJ, Sternberg K (2016) The Psychologist's companion: a guide to professional success for students, teachers, and researchers, 6th edn. Cambridge University, New York
17. Reeves M, Haanaes K, Sinha J (2015) Your strategy needs a strategy: how to choose and execute the right approach. Harvard Business Review Press, Boston
18. Kelleher JD, Tierney B (2018) Data science, ser. The MIT Press Essential Knowledge Series. MIT Press, Cambridge
19. Pearl J (2009) Causality: models, reasoning, and inference, 2nd edn. Cambridge University, New York

20. Pearl J, Glymour M, Jewell NP (2016) Causal inference in statistics: a primer. Wiley, New York
21. Imbens GW, Rubin DB (2015) Causal inference for statistics, social, and biomedical sciences: an introduction. Cambridge University, New York
22. Peters J, Janzing D, Schoelkopf B (2017) Elements of causal inference: foundations and learning algorithms. MIT Press, Cambridge

Part III
Development Opportunities of a Ph.D.

Part III presents two chapters on the development opportunities of a Ph.D.:

- **Chapter 8: Professional Development Opportunities of a Ph.D.**
- **Chapter 9: Personal Development Opportunities of a Ph.D.**

Chapter 8
Professional Development Opportunities of a Ph.D.

Chance favors the prepared mind.

—Louis Pasteur

Abstract The high intellectual standards of a Ph.D. entail substantial hard work to the extent that you may sometimes be frustrated by your Ph.D. research. Should you be less frustrated by your Ph.D. research, you can create productive time, which incentivizes you to engage with your Ph.D. research better. As you do so, you make the best out of the professional development opportunities of a Ph.D. To inspire you to seize and create professional development opportunities throughout your Ph.D. study, this chapter presents the reading, learning, writing, communication and leadership, networking, imagination, and responsible research ethics opportunities of a Ph.D.

Keywords Professional development opportunities of a Ph.D. · Seizing professional development opportunities · Creating professional development opportunities

8.1 Introduction

Anyone who stops learning is old, whether at twenty or eighty. Anyone who keeps learning stays young. The greatest thing in life is to keep your mind young.

—Henry Ford (1863–1947)

Before starting your full-fledged Ph.D. research endeavors, you must be crystal clear about the following reality of a Ph.D. study.

T. M. Getu, *Pathways for Inspired Scientists and Engineers*,
https://doi.org/10.1007/978-3-032-06784-5_8

! Attention: On Awarding a Ph.D.

A Ph.D. is awarded to a Ph.D. student who has independently carried out original research—of sufficient difficulty and scope—and contributed to one's research field while mastering the skills needed to deliver such a contribution [1].

To deliver original contributions to your Ph.D. research field, you are expected to be committed throughout your Ph.D. study, which will culminate in your prospective Ph.D. dissertation. Upon completing your doctoral study, your prospective Ph.D. dissertation is expected to deliver original contributions—of considerable difficulty and scope—to knowledge in your Ph.D. research field [2]. In this vein, your Ph.D. dissertation must demonstrate command of your research theme, your research intuition, admiration for your research discipline, your aptitude for independent research, and your ability to communicate and relate the research outcomes to the broader discourse [3, p. 2].

While committing yourself significantly for 3–5 years of your life to qualify for the high intellectual standards of a Ph.D., your Ph.D. study would offer you numerous professional development opportunities. When you swiftly seize such opportunities throughout your Ph.D. study, your likelihood of being a prospective professor increases. Nonetheless, a few professorship positions—in your chosen research field—are afforded only to a few young researchers of your generation. Consequently, it is a *cold truth* that many Ph.D. students train themselves for research jobs that may not be available throughout their future careers. In this regard, here is a question worth asking: Why do you want to pursue a Ph.D. study? This is undoubtedly one of the various questions you may be asked during your attempt to secure a Ph.D. student position. Your vision regarding this question and your unwavering commitment to realize it determine your purpose in a Ph.D. journey and the achievements of your Ph.D. study, respectively.

According to Philip J. Guo, who is the author of a memoir entitled *The Ph.D. Grind: A Ph.D. Student Memoir* [4], a Ph.D. study offers a unique environment and setup for "certain types of people" to push themselves much further than their "mental limits," so they would emerge much stronger than before [4, p. 99]. Hence, a Ph.D. study offers numerous professional and personal development opportunities. However, the same author believes such opportunities can also be cultivated without going through a Ph.D. program [4]. While this might be true with some probability under some circumstances, it can be practically infeasible to find other non-Ph.D. programs that can offer as many opportunities as a Ph.D. program, which naturally forges its trainee to not only seize the various opportunities of a Ph.D. study but also create many other opportunities. The same author also believes that some Ph.D. students may fail to seize the various professional and personal development opportunities of a Ph.D. study. This can be because some Ph.D. students are frequently breathless[1] of their Ph.D. research to the extent that they fail—with a high probability—to seize the ample opportunities afforded to Ph.D. students.

When you as a Ph.D. student are either completely absorbed or frustrated by your Ph.D. research project, your brain would manifest conscious and subconscious minds[2] that tend to either overlook or fail—with a high probability—to seize the many valuable opportunities of a Ph.D. study. Under this circumstance, your manifested conscious and subconscious minds may not be (sub-)consciously aware of the numerous ongoing Ph.D. opportunities that shall be seized daily toward not only tackling your (novel) Ph.D. research problems but also creating additional Ph.D. opportunities. On the other hand, when you are less frustrated or distracted by the natural difficulty of your Ph.D. project, you will have more focused time—at least more *neuroscientific time*[3]—to be able to exhibit conscious and subconscious minds that can make you much more productive while creating more neuroscientific time and physical time. Very interestingly, more neuroscientific and physical time will make you work harder—by motivating you regularly—so that you can engage yourself with your Ph.D. research much better. As you work harder while engaging yourself with your Ph.D. research better, you can make the best out of your Ph.D. opportunities. To this end, you would be able to not only seize the widely acknowledged opportunities of a Ph.D. study but also create additional Ph.D. opportunities, ensuring your successful transition from your Ph.D. study to your bright future. Accordingly, the number of opportunities you can seize and create during your Ph.D. study will probably determine the smoothness or roughness of your transition from the end of your Ph.D. study to your bright future, irrespective of your personal goal out of your Ph.D. study.

Regardless of your personal goal concerning your Ph.D. study, it is worth paying immense attention to all the likely opportunities of a Ph.D. study to seize and create them toward your successful Ph.D. and beyond. Thus, rather than discovering everything during your Ph.D. study and becoming surprised, we recommend that you be well aware of the numerous professional development opportunities of a Ph.D. study as early as possible. Therefore, this chapter aims to illuminate the

[1] Your scholarly journey to your Ph.D. and beyond hinges on your personal endeavor and neuroscientific capability of creating time. To this end, the time-creation opportunities of a Ph.D. study are discussed in Chap. 9, which documents the personal development opportunities of a Ph.D. study.

[2] Although there are still debates about the brain and the mind, our arguments in this chapter hinge on the view that a brain can manifest many minds (e.g., a mind for driving, a mind for eating, a mind for swimming, and so forth), pertaining to the view *the mind is a brain in action*.

[3] We define neuroscientific time as the time of one's biological neural circuits whenever one is fully concentrated on a given scholarly thinking or imagination as measured by the duration of one's manifested fully conscious or subconscious mind. This definition considers that one can concentrate on a certain scholarly thinking or imagination using a manifested fully conscious or subconscious mind (at least during the creativity phase).

many professional development opportunities[4] of a Ph.D. study, such as learning opportunities, reading opportunities, writing opportunities, communication and leadership opportunities, networking opportunities, imagination opportunities, and responsible research ethics opportunities.

The remainder of this chapter is organized as follows: Sect. 8.2 outlines the reading opportunities of a Ph.D. study. Section 8.3 presents the learning opportunities of a Ph.D. study. Section 8.4 highlights the writing opportunities of a Ph.D. study. Section 8.5 delineates the communication and leadership opportunities of a Ph.D. study. Section 8.6 details the networking opportunities of a Ph.D. study. Section 8.7 sheds light on the imagination opportunities of a Ph.D. study. Section 8.8 highlights the responsible research ethics opportunities of a Ph.D. study. Eventually, Sect. 8.9 ends this chapter with concluding remarks.

8.2 Reading Opportunities

A reader lives a thousand lives before he dies. The man who never reads lives only one.

—George R. R. Martin (1948-)

As not only the registered graduate student of your Ph.D. university but also the unregistered graduate student of all world universities, you would have abundant reading opportunities throughout your Ph.D. study. In this vein, books, monographs, book chapters, journals and magazines, conference and workshop papers, dissertations (e.g., contemplated future works), and relevant collections of images or objects are essential sources of ideas [1]. What are also crucial sources of ideas are trade magazines, popular magazines, newspapers, blogs (also vlogs), and Web sites [5, p. 147]. As for the mentioned sources of ideas, you would have the underneath reading opportunities—but not limited to—during your Ph.D. study.

! Attention: Reading Opportunities of a Ph.D. Study

- *Abundant reading opportunities in depth.*
- *Plentiful reading opportunities in breadth.*
- *Generic reading opportunities in depth and breadth.*
- *Reading on the go.*

In what follows, we highlight each of the itemized reading opportunities.

[4] Besides the professional development opportunities of a Ph.D. discussed in this chapter, the Ph.D. also offers various personal development opportunities, as detailed in Chap. 9.

8.2.1 Abundant Reading Opportunities in Depth

When you lose yourself in a book, the hours grow wings and fly.

—Chloe Thurlow (1985–)

Concerning the intricacies of your Ph.D. research field and Ph.D. research project, Ph.D. provides abundant reading opportunities in depth for the many existing books, research monographs, book chapters, journal and magazine papers, conference and workshop papers, and Ph.D. dissertations.

8.2.2 Plentiful Reading Opportunities in Breadth

There is no friend as loyal as a book.

—Ernest Hemingway (1899–1961)

Regarding the breadth of your research field that can naturally be interconnected with several other research fields, Ph.D. also provides you plentiful reading opportunities as to the numerous published books, research monographs, book chapters, journal and magazine papers, conference and workshop papers, and Ph.D. dissertations. Seizing in-breadth reading opportunities can make you envision numerous associations of concepts while advancing your synthetic and creative imagination capabilities. In doing so, you would implicitly be executing your *next-stage scholarly activities and preparedness* [6], which are paramount for your post-Ph.D. research career.

8.2.3 Generic Reading Opportunities in Depth and Breadth

Today a reader, tomorrow a leader.

—Margaret Fuller (1810–1850)

The substantial creativity and creative thinking needed—among others—throughout your Ph.D. study require you to seize generic reading opportunities both in depth and breadth. To this end, a Ph.D. study provides rich generic reading opportunities from technology magazines, trade magazines, popular magazines, newspapers, blogs, vlogs, and (academic) Web sites.

8.2.4 Reading on the Go

Reading is a discount ticket to everywhere.

—Mary Schmich (1953–)

Because leaders are always readers, your chance of creating and seizing diversified intellectual leadership opportunities will increase notably provided that you read on the go. In this vein, a Ph.D.—in tandem with twenty-first-century technologies—provides you with a chance for effective reading on the go from any researcher, world universities, research centers, research institutions, research laboratories, and alike.

Since the highlighted reading opportunities of a Ph.D. are indeed tremendous and invaluable, you shall always seize them to advance your Ph.D. research and professional development, irrespective of the challenging Ph.D. life you may be in.

Apart from the emphasized reading opportunities, a Ph.D. study also provides various learning opportunities, which are underscored next.

8.3 Learning Opportunities

You don't learn to walk by following rules. You learn by doing, and by falling over

—Richard Branson (1950–)

As a Ph.D. student, you are not only the *registered graduate student* of your enrolling university but also the *unregistered graduate student* of all universities worldwide. Thinking through this perspective, we assure you that you—as a Ph.D. student—have astronomically many opportunities for learning from not only your Ph.D. (co-)supervisor and research field researchers but also all researchers around the globe. To this end, the following learning opportunities—but not limited to—of a Ph.D. study are in order.

! Attention: Learning Opportunities of a Ph.D. Study

- *Learning in depth.*
- *Learning in breadth.*
- *Learning from universities around the world.*
- *Learning from workshops, symposiums, and conferences.*
- *Learning from colleagues and a network of researchers.*
- *Learning from the anonymous reviewers.*

- *Learning from scientific career books.*
- *Learning from self-help and social science books.*
- *Learning on the go.*

In what follows, we briefly discuss each of the itemized learning opportunities.

8.3.1 Learning in Depth

Learning never exhausts the mind.

—Leonardo da Vinci (1452–1519)

Because you, as a Ph.D. student, are expected to demonstrate vividly—among other things—mastery of your subject and new research insight [3, p. 2], a Ph.D. study provides you tremendous opportunities for persistent learning concerning the depth of your Ph.D. research field. As a Ph.D. student, you thus have abundant opportunities to learn in depth from numerous published books, monographs, book chapters, journal and magazine papers, conference and workshop papers, and Ph.D. dissertations.

8.3.2 Learning in Breadth

The mind is not a vessel to be filled but a fire to be kindled.

—Plutarch (ca. AD 40–120s)

As a Ph.D. student, you are expected to demonstrate not only mastery of your research field and new research insight but also significantly contribute to your research field by independently conducting original research [1] while demonstrating your creativity throughout your Ph.D. study. To advance your creativity and creative thinking, you are supposed to continually learn not only the depth but also the breadth of your Ph.D. research field. Accordingly, a Ph.D. study offers you ample opportunities for learning in breadth—as to your Ph.D. research field—from the various published books, monographs, book chapters, journal and magazine papers, conference and workshop papers, and Ph.D. dissertations of related and unrelated research fields.

8.3.3 Learning from Universities Around the World

Live as if you were to die tomorrow. Learn as if you were to live forever.

—Mahatma Gandhi (1869–1948)

While learning in depth and breadth of your Ph.D. research field to satisfy the mentioned Ph.D. requirements, you will have gigantic opportunities for learning from universities around the globe through their published books, monographs, book chapters, journal and magazine papers, conference papers, workshop papers, Ph.D. dissertations, and online lectures. Interestingly, the opportunities for learning from many authors of universities across the globe are remarkable, provided that you have an unwavering commitment to *learn line by line* from their innovations and publications. Accordingly, a Ph.D. study always provides abundant opportunities for learning from universities around the world, which you shall proactively seize to advance your not only Ph.D. research but also prospective research career.

8.3.4 Learning from Workshops, Symposiums, and Conferences

An investment in knowledge pays the best interest.

—Benjamin Franklin (1706–1790)

Almost all professional and learned societies regularly organize workshops, symposiums, and conferences around trending research topics or fundamental research problems. Inspiring fundamental research endeavors and trending research engagement, the regularly organized workshops, symposiums, and conferences are immense learning opportunities for any researcher and a Ph.D. student who shall always be immensely curious and open-minded. Hence, a Ph.D. study provides tremendous learning opportunities from various workshops, symposiums, and conferences, which you shall make sure you attend—throughout your Ph.D. study—with an open and inquisitive mind so that you make the best out of them.

8.3.5 Learning from Colleagues and a Network of Researchers

Learn continually. There is always 'one more thing' to learn!

—Steve Jobs (1955–2011)

In an academic program of scholarly cultivation and intellectual advancement, like a Ph.D. study, every researcher's unique experience and perspectives are always tremendous opportunities for learning, curiosity, personal growth, and professional maturity. Amid your intellectual journey to solving your unique research problems, a Ph.D. study thus provides you gigantic opportunities for learning from your colleagues enrolled at your Ph.D. university, senior and seasoned researchers of your Ph.D. research field, and a network of curious researchers whom you may wish to get connected or follow on popular academic platforms, such as *LinkedIn.*

8.3.6 Learning from the Anonymous Reviewers

I never learned from a man who agreed with me.

—Robert A. Heinlein (1907–1988)

Whenever you submit your papers to conferences and journals for peer reviews, you will come across two or more independent, anonymous reviewers who are experienced and qualified enough to evaluate the merit of your research works. The anonymous reviewers' evaluation through their criticisms, comments, and perspectives directly or indirectly indicates their subject-matter expertise, critical thinking, and mastery of your research field. Accordingly, the criticisms, comments, perspectives, and critical thinking of your prospective anonymous reviewers are ample learning opportunities to develop your capabilities in critical and creative thinking, among other essential professional skills.

8.3.7 Learning from Scientific Career Books

The more I live, the more I learn. The more I learn, the more I realize, the less I know.

—Socrates (ca. 470–399 BC)

Succeeding in an uncertain intellectual journey such as a Ph.D. study entails not only thoughtfully persistent research endeavors but also rigorous strategic thinking and an adaptive mindset. Concerning the latter, the existing invaluable books of scientific career advice—such as the ones cited by this book—put forward many invaluable professional advice and scholarly guidance on how to earn a Ph.D. and succeed as a professional researcher. As a curious and aspiring Ph.D. student, you have magnificent opportunities for proactively learning from books on scientific careers [6–18] that offer priceless professional advice and intellectual guidance, which you shall seek regularly throughout your Ph.D. study.

8.3.8 Learning from Self-Help and Social Science Books

The capacity to learn is a gift; the ability to learn is a skill; the willingness to learn is a choice.

—Brian Herbert (1947–)

As an aspiring Ph.D. student who is curious about not only your Ph.D. research but also your self-improvement, self-discovery, and bright future, you have enormous opportunities for learning philosophy, psychology, sociology, management, leadership, business, entrepreneurship, critical thinking, creative thinking, strategic thinking, probabilistic thinking, causal thinking, imaginative thinking, independent thinking, and so forth. Since it will benefit you anytime in your current or future career, you shall continually learn anything, whose direct/indirect benefit you may not know, in your personal and professional journey [3]. In this vein, you shall regularly learn from self-help and social science books by allocating time for them among your *important but non-urgent tasks and actions*. Therefore, the challenges of Ph.D. research and Ph.D. journey present *unique opportunities for learning*[5] from self-help and social science books, inspiring your self-improvement, self-discovery, and bright future.

8.3.9 Learning on the Go

I am always ready to learn although I do not always like being taught.

—Winston Churchill (1874–1965)

Because leaders are readers who are always learners, your likelihood of creating and seizing many diverse leadership opportunities will significantly increase, provided that you learn on the go. Interestingly, your ability to learn on the go—in a bus station, a train station, a bus ride, a train trip, or a plane flight—is enabled by multiple twenty first-century technologies, such as mobile Internet, smartphones, Amazon Kindles, iPads, and the like. Since performing research is a sequential thinking process, while your subconscious mind continues to work on your specific research problem, you may seize opportunities for learning on the go as opportunities for learning and thinking on the go, throughout your Ph.D. study. In learning and thinking on the go, your thinking can be inspired by your Ph.D. research problem, which further steers your learning since you have to be focused on your

[5] Your commitment to seize the crucial learning opportunities from self-help and social science books can determine your multifaceted personal development, as it will be discussed in Chap. 9.

Ph.D. research as a new researcher with limited research experience. Accordingly, a Ph.D. study offers you a chance for effective learning on the go from researchers of your research field, researchers of other related research fields, world universities, research centers, research institutes, and the like, affording a chance of learning—to any depth and breadth—anytime and anywhere, while advancing your Ph.D. research.

Because the various learning opportunities of a Ph.D.—as highlighted in Sects. 8.3.1–8.3.9—are indeed tremendous and invaluable, you shall seize them all the time, regardless of the challenging Ph.D. life you might be in while effectively managing your valuable time and creating time. Apart from those learning opportunities, a Ph.D. also provides various writing opportunities, which are emphasized next.

8.4 Writing Opportunities

Tell me and I forget, teach me and I may remember, involve me and I learn.

—Benjamin Franklin (1706–1790)

Pragmatists often describe the entire Ph.D. journey as a *publish-or-perish game*. Although we do not necessarily subscribe to such a game, we underscore that you must demonstrate—as a Ph.D. student throughout your Ph.D. study—your curiosity, critical thinking, creative thinking, and persistent hard work via a number of novel contributions that top-tier scholarly venues would publish. To this end, the following writing opportunities of a Ph.D. are in order.

! Attention: Writing Opportunities of a Ph.D. Study

- *Workshop paper writing opportunities.*
- *Conference paper writing opportunities.*
- *Journal paper writing opportunities.*
- *Magazine paper writing opportunities.*
- *Book chapter writing opportunities.*
- *Research monograph writing opportunities.*
- *Book writing opportunities.*

Underneath, each of the itemized opportunities is briefly discussed below.

8.4.1 Workshop Paper Writing Opportunities

Start writing, no matter what. The water does not flow until the faucet is turned on.

—Louis L'Amour (1908–1988)

A workshop is a mini-conference organized independently or as part of a large conference to discuss the latest advancements in a specific research theme of a given research field. As such, it can be an attractive forum for new researchers—such as new Ph.D. students—in their early stages of doctoral research. Therefore, during the initial stages of your Ph.D. research, you will have the opportunity to write and present some workshop papers under the discretion and leadership of your Ph.D. (co-)supervisor.

8.4.2 Conference Paper Writing Opportunities

Get it down. Take chances. It may be bad, but it's the only way you can do anything really good.

—William Faulkner (1897–1962)

A conference is a relatively large or very large gathering of research scholars—in academia, government, and industry—in a town/city to present and discuss the latest advancements of a given field. While relatively large conferences constitute several co-located workshops, huge ones feature many co-located symposiums on the different research themes of a cutting-edge field. As such, a Ph.D. study provides you opportunities to write and present conference papers by upgrading your research works published in workshops so that you can seize various conference participation opportunities, such as learning from, discussing with, and networking with the leading experts in your research field.

8.4.3 Journal Paper Writing Opportunities

Start before you're ready.

—Steven Pressfield (1943–)

A top-tier journal always aims to publish cutting-edge research and the latest advancements in the research field of the professional society/organization that launched it. When you get published in a top-tier journal, your article will demonstrate advanced research and cutting-edge ideas of your research field at the time of publication. Accordingly, a Ph.D. study of multi-year funded research provides you opportunities to write and publish journal papers provided that you manage to substantially advance your published conference works while demonstrating your curiosity, commitment, and persistence.

8.4.4 Magazine Paper Writing Opportunities

If you want to be a writer, you must do two things above all others: read a lot and write a lot.

—Stephen King (1947–)

A magazine is a short journal that aims to publish a given field's innovative research results, brand-new ideas, or contrasting perspectives to the state-of-the-art developments. Through a multi-year funded research in the framework of a Ph.D. project, a Ph.D. study provides you with opportunities to write and publish magazine papers, which you can seize by generating innovative research results, brand-new ideas, or novel perspectives throughout your Ph.D. study.

8.4.5 Book Chapter Writing Opportunities

You can't wait for inspiration. You have to go after it with a club.

—Jack London (1876–1916)

Book chapters are solicited by an editor(s) who is a well-known or renowned researcher(s) to publish an important edited book on a hot or emerging research field. When you conduct cutting-edge research on a hot or emerging research field, your research outputs will be attractive for publications in top-notch conferences, journals, magazines, and edited books. Accordingly, a Ph.D. study provides opportunities to write book chapters, which you can seize by carrying out bleeding-edge research in emerging or hot research fields.

8.4.6 Research Monograph Writing Opportunities

The greatest part of a writer's time is spent in reading, in order to write. A man will turn over half a library to make a book.

—Samuel Johnson (1709–1784)

When you succeed in publishing highly cited journal and magazine papers, your likelihood of publishing a research monograph—under the leadership and co-authorship of your Ph.D. (co-)supervisor—increases. Via a funded research that is sustained for 3–5 years, a Ph.D. study provides an opportunity to write a research monograph, which you shall courageously seize by publishing highly cited journal and magazine papers on your Ph.D. research field. In connection with a research monograph, a Ph.D. study also provides you a unique opportunity to write and publish an original Ph.D. dissertation, either as a research monograph or manuscript-based Ph.D. dissertation, which is a signpost on completing your Ph.D. study.

8.4.7 Book Writing Opportunities

If there's a book that you want to read, but it hasn't been written yet, then you must write it.

—Toni Morrison (1931–2019)

When it comes to a passionate Ph.D. study marked by curiosity-driven Ph.D. research of impact, the sky is the limit! Although your possibly intensive research, writing, and publication will likely make you super busy throughout your Ph.D. study, achieving groundbreaking discoveries/innovations—during this time—can pave the way to the inception of an important book project, which you can finish after earning your Ph.D. study. In this vein, a Ph.D. study provides you with a book-writing opportunity that you shall seize by devising—but not limited to—your own widely applicable novel mathematical/theoretical framework, which you are the first to discover. Meanwhile, by seizing *Peter's Laws* [19] among your success-driving strategies, you shall do it by the book as a first author—according to *Peter's Laws #12* [19]—even if your prospective book will be published only after earning your Ph.D. degree.

Besides the reading, learning, and writing opportunities stressed in Sects. 8.2, 8.3, and 8.4, respectively, a Ph.D. also provides various communication and leadership opportunities, as highlighted below.

8.5 Communication and Leadership Opportunities

Intellectual growth should commence at birth and cease only at death.

—Albert Einstein (1879–1955)

Through your scholarly and professional activities, creative thinking, critical thinking, strategic thinking, probabilistic thinking, causal thinking, and independent thinking, you can create and seize multiple communication and leadership opportunities throughout your Ph.D. study. Accordingly, the underneath are—but not limited to—crucial communication and leadership opportunities of Ph.D. study, which are worth reflecting on and seizing throughout your Ph.D. study.

! Attention: Communication and Leadership Opportunities of a Ph.D. Study

- *Giving talks.*
- *Leading from below.*
- *Forging a personality and habit of persistence.*
- *Being a multitasking professional.*
- *Volunteering for professional societies.*
- *Leading the future.*

In what follows, we highlight each of the itemized communication and leadership opportunities.

8.5.1 Giving Talks

Only the prepared speaker deserves to be confident.

—Dale Carnegie (1888–1955)

As a Ph.D. student, you are normally expected to communicate your research results in writing and speaking. Based on your research results and developing expertise in your research field, a Ph.D. study thus provides many opportunities to deliver academic talks at seminars, Webinars, workshops, symposiums, and conferences.

8.5.2 Leading from Below

Great things happen nationally when topmost leadership is goaded and supported from below.

—John W. Gardner (1912–2002)

Because a Ph.D. study entails original research that must always be carried out independently, it naturally offers various opportunities to lead from below throughout your Ph.D. study [4]. Specifically, it offers you several opportunities to lead from below through your prospective workshop and conference papers, journal and magazine papers, book chapters, and monographs.

8.5.3 Forging a Personality and Habit of Persistence

Energy and persistence conquer all things.

—Benjamin Franklin (1706–1790)

To deliver original research that you have to conduct throughout your Ph.D. study independently, you would need—among other attributes—not only sustained curiosity but also persistence to make an impact. When you work curiously and persistently by overcoming the various Ph.D. pitfalls that can happen throughout your Ph.D. study, you will likely succeed in forging a *personality and habit of persistence*. Hence, a Ph.D. study offers you multiple crucial opportunities for molding your personality and habit of persistence, which you can frequently seize to inform your personal/professional growth and professional leadership.

8.5.4 Being a Multitasking Professional

The secret to multitasking is that it isn't actually multitasking. It's just extreme focus and organization.

—Joss Whedon (1964–)

During your Ph.D. study, you may be required to revise a journal paper and prepare another conference paper—for an important conference with a looming deadline—on the same day, if not at the same hour. Such realities and challenges of a Ph.D. study shall always be seized as immense opportunities for productive multitasking[6] while molding you to be a highly effective professional. Therefore, a Ph.D. study offers you gigantic opportunities to be a multitasking professional/researcher, which you shall enthusiastically seize throughout your Ph.D. study to steer your personal/professional growth and professional leadership.

8.5.5 Volunteering for Professional Societies

There is no higher religion than human service. To work for the common good is the greatest creed.

—Woodrow Wilson (1856–1924)

Since numerous professional societies regularly organize a number of societal activities and conferences, they are always looking for passionate volunteers who

[6] Considering that the human brain can properly concentrate on only one thing at a time, some researchers advise against multitasking, which they believe is a cause for a lack of productivity. However, we believe the human brain can be trained to be a *nearly multitasking biological machine* so that it can concentrate on multiple tasks almost simultaneously, inspiring ultra-high productivity.

can be helpful. As an active and well-connected Ph.D. student, you will thus have crucial opportunities to volunteer for professional societies. When such volunteering activities are managed efficiently while being sensitive to time, they will surely be worthwhile for networking, scholarly engagement, pragmatic wisdom, professional experience, and job prospects. Accordingly, a rigorously managed Ph.D. study offers you immense opportunities for professional volunteering, which you shall seize among your *important and non-urgent tasks and actions*.

8.5.6 Leading the Future

> *Leadership is the capacity to translate vision into reality.*
>
> *—Warren Bennis (1925–2014)*

Should you learn to do research well now, you will gain a considerable advantage in the kind of research you will do later, no matter where you do it [20]. As for the Ph.D. requirements of sufficient difficulty and scope, originality, and independence [1], a Ph.D. study provides various opportunities for conducting novel research of impact. Once you manage to conduct novel Ph.D. research by seizing this opportunity, you will be poised to lead the future, at least in your research field.

On top of the communication and leadership opportunities, writing opportunities, learning opportunities, and reading opportunities briefed in Sects. 8.5, 8.4, 8.3, and 8.2, respectively, Ph.D. also presents multiple networking opportunities, as highlighted below.

8.6 Networking Opportunities

> *To realize that everything in the universe is connected is to both accept our insignificance and understand our importance in it.*
>
> *—Jeffrey Fry*

Although Ph.D. students at the early stage of their studies face an irreducible *chicken-and-egg problem* concerning networking [21], a Ph.D. study offers numerous networking opportunities. Some of these opportunities—but not limited to—are itemized below.

! Attention: Networking Opportunities of a Ph.D. Study

- *Networking for online seminars and journal clubs.*
- *Networking for scholarly experience and expertise.*
- *Networking for research collaborations.*
- *Networking for technical paper reviews.*

- *Networking for conference organizations.*
- *Networking for prospective job opportunities.*

Next, we briefly discuss each itemized networking opportunity.

8.6.1 Networking for Online Seminars and Journal Clubs

> *Networking is more about farming than it is about hunting. It's about cultivating relationships.*
>
> *—Ivan Misner (1956–)*

In the twenty first-century age of pandemics, epidemics, artificial intelligence, social networking, and rapid technological advancements, being an active Ph.D. student offers many networking opportunities for attending and actively participating in online seminars and journal clubs.

8.6.2 Networking for Scholarly Experience and Expertise

> *If you want to go somewhere, it is best to find someone who has already been there.*
>
> *—Robert T. Kiyosaki (1947–)*

In a contemporary world of *give and take* [22], you can learn from your peers while openly and passionately sharing your scholarly experience and expertise. Because a Ph.D. study equips you with unique research experience, expertise, and perspectives as to your unique Ph.D. research project, it provides ample networking opportunities for scholarly experience and expertise while openly and passionately sharing your own research experience, expertise, and perspectives. Therefore, as an active, curious, and honest Ph.D. student, you can create and seize valuable networking opportunities for sharing scholarly experience and expertise.

8.6.3 Networking for Research Collaborations

> *Networking has been cited as the number one unwritten rule of success in business. Who you know really impacts what you know.*
>
> *—Sallie Krawcheck (1964–)*

Everyone in academia, government, and industry is always on the lookout for reliable and valuable research collaborators who would engage themselves in a trusted win-win relationship. Specifically, the global research enterprise provides you invaluable networking opportunities for research collaborations, provided that you conduct and publish novel cutting-edge research that can interest a broader research community. Because a Ph.D. study provides an invaluable chance to conduct original research of impact independently, it thus offers you crucial networking opportunities for research collaborations. In this vein, it also offers you important networking opportunities for learning state-of-the-art research and development trends.

8.6.4 Networking for Technical Paper Reviews

> *Coming together is a beginning, staying together is progress, and working together is success.*
>
> *—Henry Ford (1863–1947)*

Many associate editors and editors of journals are looking for active researchers whose scholarly judgment and objectivity can facilitate their editorial leadership and services. Accordingly, connecting with some editors and associate editors would give you opportunities to provide technical paper reviews, which are also important lessons to you as an emerging scholar. Since a Ph.D. study would make you dig deeper in your field of specialization for 3–5 years, it would enable you to develop subject-matter expertise that facilitates your networking opportunities for technical paper reviews, which you can seize by being an active and curious researcher as demonstrated by your track record of publications.

8.6.5 Networking for Conference Organizations

> *Networking is not about just connecting people; it's about connecting people with people, people with ideas, and people with opportunities.*
>
> *—Michele Jennae*

Conference organizers, who are busy researchers and scholarly leaders, always look for volunteers to facilitate their conference organizations. As an active and well-published Ph.D. student, this provides you ample networking opportunities for conference organizations, which would benefit you not only with additional professional experience but also further networking opportunities. Hence, a Ph.D. study offers many networking opportunities for conference organizations, provided that you are an active, curious, and well-published Ph.D. student.

8.6.6 Networking for Prospective Job Opportunities

Everyone should build their network before they need it.

—Dave Delaney

As a curious Ph.D. student who aims to conduct bleeding-edge research, you will have ample opportunities to conduct original research and publish your results in international workshops, conferences/symposiums, and journals. As you regularly publish more workshop and conference papers as an active and curious Ph.D. student who frequently presents and takes part in several workshops and conferences, you will have abundant networking opportunities for future employment. Therefore, a Ph.D. study offers abundant networking opportunities for future job prospects as long as you frequently present and take part in several international workshops and conferences throughout your Ph.D. study.

Along with the underscored networking opportunities, communication and leadership opportunities, writing opportunities, learning opportunities, and reading opportunities highlighted in Sects. 8.6, 8.5, 8.4, 8.3, and 8.2, respectively, a Ph.D. study also presents several imagination opportunities, as highlighted below.

8.7 Imagination Opportunities

You don't understand anything until you learn it more than one way.

—Marvin Minsky (1927–2016)

The high-level prerequisite for getting your prospective manuscript published or patent granted is generating a new idea that steers your investigation on limitations, challenges, new paradigms, or new solutions concerning the state-of-the-art research advancements of your field. Because all the generated ideas cannot necessarily lead to a published manuscript or a granted patent, you shall always conduct specific detailed research regarding the ideas you have brought forth so that you can ascertain their feasibility with a high likelihood. While your endeavors of generating new ideas fall into *uplifting your imagination*, establishing their feasibility settles into *synergizing your knowledge and imagination.*

In forcing you to originate new ideas, a Ph.D. study offers countless imagination opportunities throughout your 3–5 years of Ph.D. research. These unique opportunities will uplift your *synthetic and creative imagination capabilities*, provided that you will be engaged—neuroscientifically and psychologically—in novel Ph.D. research endeavors, as demonstrated by substantial scholarly commitments of curiosity and curiosity-driven questioning during your Ph.D. study. With such scholarly commitments, you will succeed in uplifting your synthetic imagination and creative imagination capabilities, which will offer you never-ending opportunities for synergizing knowledge and imagination throughout your Ph.D. study.

Therefore, a Ph.D. study provides invaluable imagination opportunities, which are also the foundation for synergizing knowledge and imagination during your entire Ph.D. research.

Beyond the imagination opportunities, networking opportunities, communication and leadership opportunities, writing opportunities, learning opportunities, and reading opportunities discussed in Sects. 8.7, 8.6, 8.5, 8.4, 8.3, and 8.2, respectively, a Ph.D. study also offers critical opportunities for responsible research ethics, as underscored below.

8.8 Responsible Research Ethics Opportunities

The beautiful thing about learning is nobody can take it away from you.

—B.B. King (1925–2015)

Like any professional researcher, you are expected—as a Ph.D. student and new researcher—to adhere to professional ethical conduct in whatever you do [23, 24]. Specifically, you are fully responsible for the research you conduct and publish throughout your Ph.D. study. To this end, according to Peter Pruzan, who is the author of a book entitled *Research Methodology: The Aims, Practices and Ethics of Science* [23], below is the notion of responsibility that must be shouldered by all scientists—both juniors and seniors—irrespective of the kind of research they carry out [23, p. 283].

> Ethics has a demand: the scientist must assume responsibility for the way that his or her research can affect others, including nature itself. Therefore he or she must reflect not only on how results will be accepted by the scientific community, but also on how they will be accepted by all those who may be affected, directly or indirectly, by both the investigations and the results.

As for your paramount responsibility, the following are basic principles for responsible research ethics [23, 24] that are also a doctoral study's responsible research ethics opportunities, which you should abide by throughout your Ph.D. study and beyond.

! Attention: Research Ethics Opportunities of a Ph.D. Study [23, 24]

- *Never plagiarize nor engage in any act of plagiarism, including self-plagiarism.*
- *Never fabricate data.*
- *Never falsify data.*
- *Never misuse any privileged information.*
- *Never submit a conference paper to two or more conferences simultaneously.*
- *Never submit a journal article to two or more journals simultaneously.*
- *Declare upfront any of your conflict of interests.*

Besides the mentioned responsible research ethics opportunities that have been recognized by the wider research community, we also encourage you to embrace the following ethical principles that we proffer for good professional conduct.

! Attention: Basic Ethical Principles for Good Professional Conduct

- *Always give credit where it is due.*
- *Always be objective, principled, thoughtful, and trustworthy.*
- *Recognize others' endeavors as much as you want yours to be recognized.*
- *Never rush to make any judgment or conclusion.*
- *Never engage yourself in any unprofessional act and gossiping.*
- *Never hurt others with any intentional or unintentional unprofessional engagement/association.*

These principles also comprise the responsible research ethics opportunities of a Ph.D. study. Failure to embrace these opportunities can have many consequences, which will be absolutely counterproductive to your entire professional/academic journey. Therefore, we strongly encourage you to be strictly ethical in your professional/personal conduct to faithfully shoulder your important professional responsibilities as a curious emerging scholar.

8.9 Concluding Remarks

Go for it now. The future is promised to no one.

—Wayne Dyer (1940–2015)

Aiming to make you seize and create various professional development opportunities throughout your Ph.D. study, this chapter documented the professional development opportunities of a Ph.D. study. Specifically, we underscored the reading opportunities, learning opportunities, writing opportunities, communication and leadership opportunities, networking opportunities, imagination opportunities, and responsible research ethics opportunities of a Ph.D. study.

Although we attempted to discuss the most important themes regarding the professional opportunities of a Ph.D. study, our discussion of limited scope—due to a space constraint—may not have covered the *full-blown depth and breadth* of issues interwoven with the discussed topics. For elaborated insights and realities, we encourage you to reflect on this chapter's references and discuss them with your network of peers.

References

1. Mauch JE, Park N (2003) Guide to the successful thesis and dissertation: a handbook for students and faculty, 4th edn. Marcel Dekker, New York
2. Cryer P (2006) The research student's guide to success, 3rd edn. Open University Press, Berkshire
3. Petre M, Rugg G (2010) The unwritten rules of PhD research, 2nd edn. Open University Press, Berkshire
4. Guo PJ (2015) The Ph.D. Grind: a Ph.D. student memoir. http://www.pgbovine.net/PhD-memoir.htm. Cited Jan. 2020
5. Machi LA, McEvoy BT (2016) The literature review: six steps to success, 3rd edn. Corwin, Thousand Oaks
6. Reis RM (1997) Tomorrow's Professor: preparing for academic careers in science and engineering. IEEE Press, New York
7. Feibelman PJ (2011) A PhD is not enough!: a guide to survival in science, revised edn. Basic Books, New York
8. Gabrys BJ, Langdale JA (2012) How to succeed as a scientist: from postdoc to professor. Cambridge University, Cambridge
9. Rothwell N (2002) Who wants to be a scientist?: choosing science as a career. Cambridge University, New York
10. Snieder R, Larner K (2009) The art of being a scientist: a guide for graduate students and their mentors. Cambridge University Press, New York
11. Tantra R (2019) A survival guide for research scientists. Springer, Cham
12. Ramón y Cajal S (1999) Advice for a young investigator (N. Swanson and L. W. Swanson, Trans.). The MIT Press, Cambridge
13. Medawar PB (1979) Advice to a young scientist. Harper & Row, New York
14. Rosei F, Johnston T (2006) Survival skills for scientists. Imperial College Press, London
15. Shneiderman B (2016) The new ABCs of research: achieving breakthrough collaborations. Oxford University Press, Oxford (Kindle Edition)
16. Olson JS, Olson GM (2013) Working together apart: collaboration over the internet. Morgan & Claypool, New York
17. Sternberg RJ, Sternberg K (2016) The Psychologist's companion: a guide to professional success for students, teachers, and researchers, 6th edn. Cambridge University, New York
18. Ghezzi C (2020) Being a researcher: an informatics perspective. Springer, Cham
19. Diamandis PH (2024) Peter's Laws. https://www.diamandis.com/peters-laws
20. Booth WC, Colomb GG, Williams JM, Bizup J, FitzGerald WT (2016) The craft of research, 4th edn. The University of Chicago Press, Chicago
21. Agre P (2002) Networking on the network: a guide to professional skills for PhD students, 2nd edn. University of California—Los Angeles, Los Angeles
22. Grant A (2013) Give and take: a revolutionary approach to success. Viking, New York
23. Pruzan P (2016) Research methodology: the aims. Practices and Ethics of Science, Springer
24. National Academy of Sciences (2009) On being a scientist: a guide to responsible conduct in research, 3rd edn. The National Academies Press, Washington D.C.

Chapter 9
Personal Development Opportunities of a Ph.D.

> *Personal development – the never ending chance to improve not only yourself but also attract opportunities and affect others.*
>
> *—Jim Rohn (1930–2009)*

Abstract Many thinkers agree that personal development is a never-ending critical endeavor that takes time, energy, willingness, and persistence. By the same token, your commitment to earning your Ph.D. is an enlightening personal development expedition whose success hinges on your ability to seize and create various personal development opportunities throughout your Ph.D. study. Accordingly, we present the service, time creation, leadership, thinker development, shaping and being yourself, and miscellaneous opportunities of a Ph.D. study.

Keywords Personal development opportunities of a Ph.D. · Seizing personal development opportunities · Creating personal development opportunities

9.1 Introduction

> *Personal development is the belief that you are worth the effort, time, and energy needed to develop yourself.*
>
> *—Denis Waitley (1933–)*

It is famously attributed to the Buddha [1]—the founder of *Buddhism*—that *we become what we think*. Our thinking, essentially determining our identity (or persona), indeed constitutes the fundamental neuroscientific phenomena that have made each of us unique. In encoding the unique yourself, the remarkably mysterious neuroscientific phenomena—the subject of various ongoing neuroscientific

T. M. Getu, *Pathways for Inspired Scientists and Engineers*,
https://doi.org/10.1007/978-3-032-06784-5_9

investigations[1]—are widely believed to pave the way to your behavior, cognitive and motor skills, knowledge, awareness, intelligence, and alike. Producing these attributes through multifaceted neuroscientific developments encompasses personal development. Therefore, from the vantage point of neuroscience, your personal development is your endeavor of continuously wiring and re-wiring your brain/mind to create the best version of yourself.

Toward self-actualizing yourself, the continual wiring and re-wiring of your mind can be predominantly affected by your overall real-world life experience; your environment that you would spend your most time at; and your *intellectual neural experience* (i.e., cognitive experience) such as your disposition to think critically, creatively, strategically, probabilistically, causally, and independently. Such cognitive endeavors take—as also quoted above for Denis Waitley—considerable time, energy, willingness, and persistence. As an interesting association, these are also what it would take to succeed in your intellectual journey toward your Ph.D. and beyond. Hence, your persistent undertaking of committing yourself to qualify for your novel Ph.D. is an illuminating personal development expedition.

Born out of your successful expedition of personal development, your significant personal development often informs and guides your professional development, which in turn contributes to the further advancement of your personal development. Therefore, your personal development is a compass guide to your bright future, which can be achieved through a settled professional or research career. Contrarily to such bright prospects afforded to you through your uninterrupted personal development, a lack of personal development darkens—with a high probability—your future and career prospects. Specifically, as agreed by many thinkers, pushing aside your personal development amounts to inviting personal struggle to your professional and personal lives. In this vein, your probability of struggling in your Ph.D. study significantly increases when you frequently brush off on your personal development endeavors. When you brush off on essential personal development, you will yield your hampered professional career. Your hampered professional career—continuing with our cause-and-effect analysis—will then be a major time-waster for your Ph.D. study and your personal endeavor beyond your Ph.D. study.

According to Brian Tracy [2], who is a Canadian-American motivational speaker and self-development author, your personal development is your major time-saver: The better you get, the less time it will take you to achieve your goals.[2] On the other hand, according to Warren Buffet, who is an American business magnate and one of the most successful investors in the world [3], the best investment you can make is

[1] Regarding the fundamental neuroscientific phenomena that pave the way to our personal development, there are intertwined molecular realities, cellular realities, neuronal network realities, neuronal systems realities, cognitive realities, and behavioral realities, which are the overarching research themes of molecular neuroscience, cellular neuroscience, network neuroscience, systems neuroscience, cognitive neuroscience, and behavioral neuroscience, respectively.

[2] "Personal development is a major time-saver. The better you become, the less time it takes you to achieve your goals."—Brian Tracy (1944-).

in yourself.[3] In this regard, your frequent personal development endeavor is the best investment you can make in light of your not only intellectual journey to a Ph.D. but also your professional/scholarly stride beyond a Ph.D. study. Accordingly, you shall not only seize opportunities for personal development throughout your Ph.D. study but also work tirelessly and thoughtfully to create additional opportunities for personal development.

As a Ph.D. student, your expedition of enlightened personal development hinges upon your unwavering endeavors of seizing the various personal development opportunities of a Ph.D. study while also creating additional personal development opportunities throughout your Ph.D. study. Aiming to inspire your successful Ph.D. study, which shall also be an immense opportunity for your enlightened personal development, we hereby present the various personal development opportunities of a Ph.D. study. Specifically, we underscore the service opportunities of a Ph.D. study, the time-creation opportunities of a Ph.D. study, the leadership opportunities of a Ph.D. study, the thinker development opportunities of a Ph.D. study, the Ph.D. study opportunities of shaping and being yourself, and the miscellaneous opportunities of a Ph.D. study. All these opportunities can inspire and empower you to not only seize all the personal development opportunities afforded by your Ph.D. study but also create many additional personal development opportunities for your bright professional or scholarly future.

The remainder of this chapter is organized as follows. Section 9.2 presents the service opportunities of a Ph.D. study. Section 9.3 reveals the time-creation opportunities of a Ph.D. study. Section 9.4 details the leadership opportunities of a Ph.D. study. Section 9.5 enlightens the thinker development opportunities of a Ph.D. study. Section 9.6 describes the Ph.D. study opportunities of shaping and being yourself. Section 9.7 affirms the miscellaneous opportunities of a Ph.D. study. Finally, Sect. 9.8 ends this chapter with concluding remarks.

9.2 Service Opportunities

> *The best way to find yourself is to lose yourself in the service to others.*
>
> *—Mahatma Gandhi (1869–1948)*

A Ph.D. study offers several professional service opportunities, which are also opportunities for personal development throughout your years of impactful research. Specifically, the following are—but not limited to—some of the crucial service opportunities of a Ph.D. study that inspire your continued personal and professional development, provided that you seize them effectively as they arise.

[3] "By far the best investment you can make is in yourself."—Warren Buffett (1930-).

! Attention: Service Opportunities of a Ph.D. Study

- *Workshop organization.*
- *Conference/symposium organization.*
- *Technical review of workshop papers.*
- *Technical review of conference/symposium papers.*
- *Technical review of journal papers.*

Next, we highlight each itemized service opportunity of a Ph.D. study.

9.2.1 Workshop Organization

Our rewards in life will always be in direct ratio to our service.

—Earl Nightingale (1921–1989)

Workshop organizers—who are often very busy researchers—are usually looking for volunteers who can facilitate their workshop organization. If you manage to be a visibly active Ph.D. student who is enthusiastic and well connected, a Ph.D. study provides you a number of workshop organization opportunities, which you shall regularly seize to make the best out of them and advance your personal development.

9.2.2 Conference/Symposium Organization

Those who are happiest are those who do the most for others.

—Booker T. Washington (1856–1915)

Being particularly very busy researchers, conference/symposium organizers are often looking for volunteers who can meaningfully contribute to the success of their organized conferences/symposiums. If you manage to be a very active Ph.D. student who is a passionate, well-connected, and new researcher, conference/symposium organization opportunities will often pop up in your professional way. By seizing these opportunities and making the best out of them, you can advance your professional and personal development while seizing other opportunities for a Ph.D. study.

9.2.3 Technical Review of Workshop Papers

Everyone can be great because everyone can serve.

—Martin Luther King, Jr. (1929–1968)

If you manage to be an active, passionate, productive, and well-connected Ph.D. student, you will likely receive multiple professional invitations for the technical review of workshop papers on your research field. Accordingly, a Ph.D. study provides multiple opportunities for the technical review of workshop papers, which you shall seize to not only serve other authors—in constructively criticizing their papers and providing feedback on their work—but also further learning the state-of-the-art research developments of your research field.

9.2.4 Technical Review of Conference/Symposium Papers

The more we give away, the more is given to us.

—Wayne W. Dyer (1940–2015)

If you manage to be an active, productive, and well-connected Ph.D. student, you will also receive many invitations for the technical review of conference/symposium papers that attempt to address problems in your research field. Thus, a Ph.D. study provides multiple opportunities for the technical review of conference/symposium papers, which you shall seize to not only help others—in providing constructive criticisms and feedback on their papers—but also further study the bleeding-edge research advancements of your research field.

9.2.5 Technical Review of Journal Papers

When we give cheerfully and accept gratefully, everyone is blessed.

—Maya Angelou (1928–2014)

Because many editors and associate editors are looking for active researchers whose expertise, scholarly judgment, and objectivity can facilitate their editorial services and leadership, multiple journal paper review opportunities can pop up, provided that you are an active, productive, and well-connected Ph.D. student. Therefore, a Ph.D. study provides multiple opportunities for the technical review of journal papers, which you shall always seize to not only support others—in offering constructive criticisms and feedback on their journal articles—but also further learn the cutting-edge research developments of your research field while developing yourself professionally and personally.

Besides the highlighted service opportunities, a Ph.D. study also offers many time-creation opportunities—among its personal development opportunities—which are discussed next.

9.3 Time-Creation Opportunities

Time stays long enough for anyone who will use it.

—Leonardo da Vinci (1452–1519)

The success of any researcher—whether senior or junior—depends mainly on one's ability to create time while sensibly engaging for valuable time. As we alluded to in the preceding chapter, time creation corresponds to the creation of not only physical time (i.e., conventional time) but also *neuroscientific time*. As defined in Chap. 8, neuroscientific time is the time of one's biological neural circuits whenever one is fully concentrated on a given scholarly thinking or imagination as quantified by the duration of one's manifested fully conscious or subconscious mind. By entirely concentrating on a given scholarly thinking or imagination, you can always create ample neuroscientific time, irrespective of the elapsed physical time. Accordingly, you can create both physical and neuroscientific time[4]—during your Ph.D. study—by exploiting different time creation strategies.

If a Ph.D. study is done right, it will forge you into an effective time manager who can often create both physical time and neuroscientific time using various strategies, which are also the time-creation opportunities of a Ph.D. study. Some of these opportunities are—but not limited to—itemized below.

! Attention: Time-Creation Opportunities of a Ph.D. Study

- *Time creation via high effectiveness.*
- *Time creation through high performance.*
- *Time creation via hyperfocus.*
- *Time creation through research collaboration.*
- *Time creation via motivation.*
- *Time creation through depth and breadth of thinking.*
- *Time creation via imagination.*
- *Time creation through vision.*

In what follows, we highlight each itemized opportunity for time creation.

[4] Despite the beautiful and inspiring contributions of the multidisciplinary neuroscience research communities, we may not have a fundamental understanding of how neuroscientific time is created.

9.3.1 Time Creation via High Effectiveness

You will never find time for anything. If you want time you must make it.

—Charles Buxton (1822–1871)

When you become a highly effective researcher, you can create much more time by effectively executing not only *urgent and important intellectual tasks and actions* but also *important and not urgent intellectual tasks and actions* critical for your personal and professional growth. To successfully create time using this strategy, you shall be focused on figuring out important and not urgent intellectual tasks and actions—since many things may be hidden in a busy research world—that you would invest your time on to create much more valuable time. Accordingly, a Ph.D. study offers you many opportunities to create time through high effectiveness by steering you to effectively engage with various intellectual tasks and actions that are not only urgent and important but also important and not urgent.

9.3.2 Time Creation through High Performance

Time is a created thing. To say "I don't have time," is like saying, "I don't want to."

—Lao Tzu (b. 571 BC)

Embracing the habits of high-performing people can make you create time through their personal and social habits [4].[5] The high-performing people's habits that empower you to create time are *seeking clarity, generating energy, and raising necessity*; the social habits that enable you to create time are *increasing productivity, developing influence, and demonstrating courage* [4, p. 39–186]. Consequently, a Ph.D. study offers you ample opportunities to create time through high performance while forging you to generate energy, seek the clarity of your study, raise the necessity of your work, demonstrate scientific courage, step up the productivity of your work, and develop scientific influence inside and outside your research community.

[5] The book in [4] is a self-help book whose author has put decades of research into it. Although the academic community may not respect self-help books because of their insufficient evidence, we refer to [4] since it is a valuable resource featuring invaluable advice for Ph.D. students.

9.3.3 Time Creation via Hyperfocus

Concentrate all your thoughts upon the work at hand. The sun's rays do not burn until brought to a focus.

—Alexander Graham Bell (1847–1922)

Hyperfocus is an accruing mind state when someone is in an intense mental concentration and visualization that profoundly focuses one's consciousness on a subject, task, or topic [5]. In some people, several topics or subjects may include fiction, concepts, objects of the mind, and imagination [5]. Through intense mind concentration and visualization, hyperfocus can make you create ample time by enabling you to complete several tasks quickly while completely absorbed in what you do. While hyperfocus can be a cognitive condition that can sometimes be detrimental in some real-world situations, it can be an indispensable strategy for creating time, especially in a distracted twenty first-century world of social media and a deluge of information. Hence, a Ph.D. study offers many opportunities to create abundant time through hyperfocus, especially neuroscientific time.

9.3.4 Time Creation Through Research Collaboration

There is no such thing as a self-made man. You will reach your goals only with the help of others.

—George Shinn (1941–)

A productive research collaboration can make you create plentiful time by leveraging the expertise and guidance of your research collaborators. However, too much collaboration may not be appropriate for you as a Ph.D. student since you have yet to demonstrate your ability to perform original research independently. Although a productive collaboration should first be encouraged by your Ph.D. (co-)supervisor, who would normally expect you to be fully independent all the time, a Ph.D. study provides you valuable collaboration opportunities—with senior and accomplished researchers—that undoubtedly help you create abundant time during your Ph.D. study.

9.3.5 Time Creation via Motivation

The bad news is time flies. The good news is you're the pilot.

—Michael Altshuler

Possibly sparked by your *ambition and expectancy*, your intrinsic motivation often makes you create time by empowering you to persistently seek novel results, irrespective of the hurdles you may encounter. As your intrinsic motivation is sustained through regular attention and effort and then amplified via the right attitude and environment, you will undoubtedly create ample time. The invaluable time you would create using motivation is copious neuroscientific time because sustained and amplified motivation steers you to be laser focused on your Ph.D. research while disincentivizing you to be distracted by any challenge that can easily be frustrating. This is precisely how a Ph.D. study rewards you for being motivated and focused throughout your years of doctoral research, enabling you to create ample physical and neuroscientific time through motivation.

9.3.6 Time Creation Through Depth and Breadth of Thinking

> *Everybody has 24 hours and the question is, what do you do with your 24 hours? That's what makes everybody equal.*
>
> *—Stedman Graham (1951–)*

When you independently think about the advancements of your research field in depth and breath, you will be able to foresee not only the most crucial intricacies and challenges of your research field but also the many big-picture issues and challenges at the heart of your research field, paving the way for creating intellectual opportunities of innovation. Thinking in depth and breadth can hence make you create time through your developing depth of expertise that would enable you to anticipate important details and challenges of the research problems you may wish to attack; it also makes you foreknow the paramount big-picture issues—concerning your undertaken research problems—through your emerging breadth of expertise. By engaging yourself with depth and breadth of thinking, you can thus create plentiful neuroscientific time that would likely pave the way for your important innovation and impact. Accordingly, a Ph.D. study provides you with opportunities to think in depth and breadth, allowing you to create sufficient neuroscientific time throughout your Ph.D. study.

9.3.7 Time Creation via Imagination

> *Imagination rules the world.*
>
> *—Napoleon Bonaparte (1769–1821)*

Imagination naturally makes you create time—both neuroscientific and physical time—by enabling you to generate new ideas while deploying advanced skills in *synthetic and creative imagination*. Through the faculty of synthetic imagination, you can organize existing ideas and concepts into new combinations while merely exploiting your observation, education, and experience [6].[6] Through the faculty of creative imagination, you can generate basic or new ideas while entertaining direct communication with "infinite intelligence" [6]. Therefore, since a Ph.D. study would forge you to acquire advanced skills in synthetic and creative imagination, it would offer you many opportunities for creating abundant physical and neuroscientific time.

9.3.8 Time Creation Through Vision

> *I have a secret project which adds four hours every day to the 24 hours we have. There's a bit of time travel involved.*
>
> *—Sundar Pichai (1972–)*

A research vision is an *intellectual GPS (Global Positioning System)* that guides you where you want to go and what you can accomplish in science and innovation. With a research vision, you would know where you are going and what you can likely accomplish; with no research vision, you would unlikely know where you are going, and you will likely achieve nothing significant, regardless of your endeavor. Hence, a novel research vision determines not only your next destination but also your nearly certain upcoming chance of success while saving you much more time that would otherwise be spent intellectually exploring without a research vision.

Besides saving you ample time, an innovative research vision naturally enables you to create *abundant time toward the future*, provided that you envision novel interdisciplinary, multidisciplinary, and transdisciplinary research, which are essential for the artificial intelligence (AI)-era of the twenty first century. Accordingly, because a Ph.D. research in the AI-driven technological era engages you with not only disciplinary research but also interdisciplinary, multidisciplinary, and transdisciplinary *convergence research* [7, 8], a Ph.D. study provides many opportunities to create plentiful time through an innovative vision. For the same reason, a Ph.D. study also provides many opportunities to create abundant time toward the future through interdisciplinary, multidisciplinary, and transdisciplinary research that is visionary and bold.

Besides the time creation and service opportunities highlighted in Sects. 9.3 and 9.2, respectively, a Ph.D. study also offers various leadership opportunities, as underscored below.

---

[6] The classic book in [6] is a self-help book whose author has conducted decades of research on it. Although the academic community may not respect self-help books because of their insufficient evidence, we refer to [6] because it encompasses invaluable advice for Ph.D. students.

9.4 Leadership Opportunities

Leadership and learning are indispensable to each other.

—John F. Kennedy (1917–1963)

As partly discussed in Sect. 8.5, which concerns the communication and leadership opportunities of a Ph.D. study in light of the professional opportunities of a Ph.D. study, a Ph.D. study offers you various leadership opportunities that are also immense opportunities for personal development. Some of these opportunities—but not limited to—are itemized below.

! Attention: Leadership Opportunities of a Ph.D. Study

- *Leading from below.*
- *Leading in depth.*
- *Leading in breadth.*
- *Leadership through an event organization.*
- *Leading the future.*

Next, we briefly discuss each itemized leadership opportunity.

9.4.1 Leading from Below

Innovation distinguishes between a leader and a follower.

—Steve Jobs (1955–2011)

As highlighted before, a Ph.D. study offers you multiple opportunities to lead from below [9]—under the leadership of your Ph.D. (co-)supervisor—through your prospective workshop papers, conference papers, journal articles, magazine articles, book chapters, and monographs.

9.4.2 Leading in Depth

A leader is one who knows the way, goes the way, and shows the way.

—John C. Maxwell (1947–)

While thoughtfully, courageously, and persistently leading from below throughout your Ph.D. study, a Ph.D. study also provides you with many opportunities to lead in depth, which you can surely seize by attacking the depth of research problems regarding your Ph.D. research project.

9.4.3 Leading in Breadth

If your actions inspire others to dream more, learn more, do more and become more, you are a leader.

—John Quincy Adams (1767–1848)

While courageously, thoughtfully, and persistently leading from below throughout your Ph.D. study, a Ph.D. study also gives you valuable opportunities to lead in breadth, which you can certainly seize by uncovering big-picture issues touching upon your research problems emanating from your Ph.D. research project. Although leading in depth is the natural opportunity of a Ph.D. study, since you may work on a few specific research themes throughout your Ph.D. study, you can also lead in breadth by exposing the significant big-picture issues associated with the research problems that you would formulate in line with your Ph.D. project.

9.4.4 Leadership Through an Event Organization

Life's most persistent and urgent question is, "What are you doing for others?"

—Martin Luther King Jr. (1929–1968)

When you manage to be an active, visibly published, and well-connected Ph.D. student, a Ph.D. study provides you with unique opportunities of leadership—under the leadership of your Ph.D. (co-)supervisor and other leading researchers—in jointly organizing webinars, seminars, workshops, conferences/symposiums, and regular professional meetings, which are all invaluable opportunities for your strategic networking and post-Ph.D. career prospects.

9.4.5 Leading the Future

Go as far as you can see; when you get there, you'll be able to see farther.

—J. P. Morgan (1837–1913)

So long as you are committed, persistent, and courageous enough to be a visionary researcher throughout your Ph.D. study, a Ph.D. study offers numerous opportunities to lead the future. To seize these priceless opportunities and succeed in leading the future, you may consistently demonstrate visionary leadership throughout your Ph.D. study while leading from below, in depth, and in breadth. By doing so, you can successfully create plentiful physical and neuroscientific time that would empower your scientific/technological leadership of the future, at least regarding your field.

Besides the leadership opportunities, time-creation opportunities, and service opportunities highlighted in Sects. 9.4, 9.3, and 9.2, respectively, a Ph.D. study also offers you many thinker development opportunities, as underlined below.

9.5 Thinker Development Opportunities

> *Any ideas, plan, or purpose may be placed in the mind through repetition of thought.*
>
> *—Napoleon Hill (1883–1970)*

Thinker development opportunities are the most important among the numerous personal development opportunities of a Ph.D. study because anything can easily be accomplished once you have successfully developed yourself as an independent thinker. Due to the many inevitable Ph.D. pitfalls; the demanding nature of being a new researcher in the uncharted intellectual landscape; the natural uncertainty associated with conducting original research; and the high intellectual standards of a doctoral study, a Ph.D. study offers you numerous personal development opportunities that forge you to develop as an emerging thinker provided that it is done rigorously with a singular purpose. Some of these development opportunities—but not limited to—are listed below.

! Attention: Thinker Development Opportunities of a Ph.D. Study

- *Critical thinking opportunities.*
- *Creative thinking opportunities.*
- *Strategic thinking opportunities.*
- *Probabilistic thinking opportunities.*
- *Causal thinking opportunities.*
- *Independent thinking opportunities.*

In what follows, we briefly discuss each itemized development opportunity.

9.5.1 Critical Thinking Opportunities

The important thing is not to stop questioning.

—Albert Einstein (1879–1955)

In attempting to overcome the unavoidable research-related Ph.D. pitfalls, writing-related Ph.D. pitfalls, publication-related Ph.D. pitfalls, management-related Ph.D. pitfalls, persona-related Ph.D. pitfalls, motivation-related Ph.D. pitfalls, and generic Ph.D. pitfalls, you would have innumerably many diversified opportunities of developing as a critical thinker. What also offers you opportunities for developing as a critical thinker are the tradeoffs of a Ph.D. study; the uncertain nature of an intellectual journey; the rapidly evolving research landscape of the twenty first-century science, technology, engineering, and mathematics (STEM); the ultra-competitive nature of the contemporary research enterprise; and so forth. Because a modern Ph.D. study affords all these opportunities, it undoubtedly offers you many diverse opportunities that are invaluable for developing as a critical thinker.

9.5.2 Creative Thinking Opportunities

Creativity involves breaking out of established patterns in order to look at things in a different way.

—Edward de Bono (1933–2021)

In aiming to surmount the inevitable research-related Ph.D. pitfalls, writing-related Ph.D. pitfalls, publication-related Ph.D. pitfalls, management-related Ph.D. pitfalls, persona-related Ph.D. pitfalls, motivation-related Ph.D. pitfalls, and generic Ph.D. pitfalls, you would have numerous diversified opportunities of developing as a creative thinker while continuing to engage yourself in creative thinking that can mitigate the detrimental impact of the mentioned Ph.D. pitfalls. You would also have many invaluable opportunities to develop as a creative thinker when you creatively think through the rapidly changing research landscape of the twenty first-century STEM; seize the ultra-competitive nature of the contemporary research enterprise as an ample opportunity for creativity; routinely engage yourself in non-conformist thinking; aim for convergent and divergent thinking aimed at (unconventional) creativity; frequently pursue synthetic and creative imagination; and so forth. Since a contemporary Ph.D. study affords all these opportunities, it certainly provides numerous diverse opportunities for your development as a creative thinker.

9.5.3 Strategic Thinking Opportunities

Strategy is about making choices, trade-offs; it's about deliberately choosing to be different.

—Michael Porter (1947–)

In targeting to overcome the many inevitable research-related Ph.D. pitfalls, writing-related Ph.D. pitfalls, publication-related Ph.D. pitfalls, management-related Ph.D. pitfalls, persona-related Ph.D. pitfalls, motivation-related Ph.D. pitfalls, and generic Ph.D. pitfalls, you would have innumerably many diverse opportunities of developing as a strategic thinker, while periodically engaging yourself in strategic thinking that can lessen the detrimental impact of the mentioned Ph.D. pitfalls. You would also have ample opportunities to evolve as a strategic thinker when you strategically think through the tradeoffs of a Ph.D. study; strategically think through the rapidly changing research landscape of the twenty first-century STEM; seize the ultra-competitive nature of the contemporary research enterprise as a gigantic opportunity for strategic thinking; strategically think through the uncertain nature of an intellectual journey; regularly think through the vantage point of a strategy for a strategy; keep accumulating pragmatic experience, knowledge, and wisdom that enable you to think through the lens of strategy; and so forth. Considering that a contemporary Ph.D. study affords almost all these opportunities, it surely provides many diverse opportunities invaluable for your development as a strategic thinker.

9.5.4 Probabilistic Thinking Opportunities

The most important questions of life are indeed, for the most part, really only problems of probability.

—Pierre-Simon Laplace (1749–1827)

In aspiring to surmount the many inevitable research-related Ph.D. pitfalls, writing-related Ph.D. pitfalls, publication-related Ph.D. pitfalls, management-related Ph.D. pitfalls, persona-related Ph.D. pitfalls, motivation-related Ph.D. pitfalls, and generic Ph.D. pitfalls, you would have various opportunities of developing as a probabilistic thinker while frequently engaging yourself in probabilistic thinking that can diminish the detrimental impact of the mentioned Ph.D. pitfalls. You would also have plentiful opportunities to develop as a probabilistic thinker when you probabilistically think through the tradeoffs of a Ph.D. study; seize the ultra-competitive nature of the contemporary research enterprise as an immense opportunity for probabilistic thinking; acquire pragmatic experience, knowledge, and wisdom to think through the lens of probability; probabilistically think through the uncertain nature of an intellectual journey; probabilistically think through the rapidly changing research landscape of the twenty first-century STEM; gain probabilistic wisdom to think intensively using probability as an overarching lens; and so forth. Because a contemporary Ph.D. study affords almost all these opportunities, it certainly provides numerous diverse opportunities that are priceless for your development as a probabilistic thinker.

9.5.5 Causal Thinking Opportunities

I believe that whatever we do or live for has its causality; it is good, however, that we cannot see through to it.

—Albert Einstein (1879–1955)

In intending to overcome the many inevitable research-related Ph.D. pitfalls, writing-related Ph.D. pitfalls, publication-related Ph.D. pitfalls, management-related Ph.D. pitfalls, persona-related Ph.D. pitfalls, motivation-related Ph.D. pitfalls, and generic Ph.D. pitfalls, you would have many diverse opportunities of developing as a causal thinker, while routinely engaging yourself in causal thinking that can minimize the detrimental impact of the named Ph.D. pitfalls. You would also have abundant opportunities to develop as a causal thinker when you causally think through the tradeoffs of a Ph.D. study; causally think through the rapidly changing research landscape of the twenty first-century STEM; causally think through the uncertain nature of an intellectual journey; amass significant scholarly experience in critical thinking, creative thinking, strategic thinking, and probabilistic thinking that can accelerate your development as a causal thinker; gain significant pragmatic experience, knowledge, and wisdom to think through the lens of causality; acquire magnificent causal wisdom to think deeply using causality as a primary lens; and so on. Since a contemporary Ph.D. study affords almost all these opportunities, it provides numerous diverse opportunities invaluable for your development as a causal thinker.

9.5.6 Independent Thinking Opportunities

Life isn't about finding yourself. Life is about creating yourself.

—George Bernard Shaw (1856–1950)

In committing yourself to overcome the many inevitable research-related Ph.D. pitfalls, writing-related Ph.D. pitfalls, publication-related Ph.D. pitfalls, management-related Ph.D. pitfalls, persona-related Ph.D. pitfalls, motivation-related Ph.D. pitfalls, and generic Ph.D. pitfalls, you would have many diverse opportunities of evolving as an independent thinker while routinely engaging yourself in independent thinking that can minimize the detrimental impact of the named Ph.D. pitfalls. You would also have ample opportunities to develop as an independent thinker when you independently and intensively think through the rapidly changing research landscape of the twenty first-century STEM; independently and intensively think through the uncertain nature of an intellectual journey; garner considerable scholarly experience in critical thinking, creative thinking, strategic thinking, probabilistic thinking, and causal thinking that inspire your development as an independent thinker; acquire significant pragmatic experience, depth and breadth of expertise, and sage wisdom to think independently and intensively; gain strategic wisdom, probabilistic wisdom, and causal wisdom to think independently and intensively; and so on. Because a contemporary Ph.D. study affords almost all these opportunities, it undoubtedly provides many diverse opportunities that are priceless for your development as an independent thinker.

Apart from the thinker development opportunities, the leadership opportunities, the time-creation opportunities, and the service opportunities discussed in Sect. 9.5, 9.4, 9.3, and 9.2, respectively, a Ph.D. study also provides you various opportunities to shape and be yourself, as argued below.

9.6 Opportunities of Shaping and Being Yourself

By far the best investment you can make is in yourself.

—Warren Buffett (1930–)

A successful and engaging Ph.D. journey by itself offers you numerous opportunities that would certainly inspire you to not only shape yourself but also be yourself, in any of your prospective walks of professional/personal life. This is because a Ph.D. study neuroscientifically immerses you in unique intensive thinking while you persevere in executing new challenging intellectual opportunities for 3–5 years, forging the unique you who usually have satisfaction in being yourself. Some of these opportunities—but not limited to—are itemized below.

! Attention: Ph.D. Study Opportunities of Shaping and Being Yourself

- *Developing your writing style.*
- *Forging your thinking style.*
- *Developing your reasoning style.*
- *Cultivating your research and curiosity style.*

In what follows, we highlight each of the itemized opportunities.

9.6.1 Developing Your Writing Style

A writer is someone for whom writing is more difficult than it is for other people.

—Thomas Mann (1875–1955)

Among the requirements of a Ph.D., you must deliver original research, which you would demonstrate by publishing top-tier journal papers. In doing so, a Ph.D. study offers you numerous opportunities to develop your writing/authoring style that you can seize by choosing your preferred words, sentence structure and length, metaphors, rhythm and pacing, tone, distinctive patterns/habits, and so forth while inspired by the elegant styles and prose of exemplary great writers.

9.6.2 Forging Your Thinking Style

If everyone is thinking alike, then somebody isn't thinking.

—George S. Patton (1885–1945)

The Ph.D. requirements of originality and complete independence entail persistent hard work and intensive thinking throughout your Ph.D. study. To this end, a Ph.D. study provides ample opportunities for thinking throughout your Ph.D. study, irrespective of the research problems you would be working on. Specifically, you would learn to think in depth and breadth of issues/perspectives; to think through not only your perspectives but also others' viewpoints; how to be a critical thinker, a creative thinker, a strategic thinker, a probabilistic thinker, and a causal thinker—all for yourself by yourself; the many ways that can pave the way to being an independent thinker not only as a Ph.D. student but also an independent researcher; and so forth.

9.6.3 Developing Your Reasoning Style

Absorb what is useful. Discard what is not. Add what is uniquely your own.

—Bruce Lee (1940–1973)

For 3–5 years, a Ph.D. study is always a personal period of uninterrupted learning and discovery toward a novel Ph.D. dissertation that would report on your independently conducted series of original works. During this period of learning, intellectual growth, and enriching experience, the many papers you would be reading, analyzing, and comprehending; the various conference/journal papers you would be writing and publishing; and the diverse papers you would be reviewing as an anonymous technical reviewer will undoubtedly teach you a lot about reasoning, such as inductive reasoning, deductive reasoning, and abductive (plausible) reasoning, or a combination of them. Because the published authors' papers feature one or more of these reasoning types and their variants that can all forge you to acquire your reasoning style, a Ph.D. study offers you plentiful opportunities for developing your reasoning style, paving the way to your advanced reasoning skills.

9.6.4 Cultivating Your Research and Curiosity Style

I want to sing like the birds sing, not worrying about who hears or what they think.

—Rumi (1207–1273)

A Ph.D. study offers you numerous opportunities to develop your research and curiosity styles—throughout your Ph.D. study—under the guidance and close collaboration of your Ph.D. (co-)supervisor. While the experience and advice of your Ph.D. (co-)supervisor on conducting novel research are always important, there is no silver bullet research and curiosity style that would make you successful in your endeavors of research and curiosity all the time. Since novel research is always a dynamic and challenging intellectual endeavor, conducting novel research is like driving in a new big city—with too many routes and traffic—without any map. The absence of any map implies that you would be intellectually sailing in a new territory that you do not have any prior knowledge of, starting from what you know or perceive to know; on the other hand, the presence of too much traffic indicates that you would have too many competitors—especially in these days—in advancing your preferred research and curiosity for an impactful discovery/innovation. Given an impactful discovery/innovation, a probably pragmatic strategy would be getting involved in novel research while being courageous enough to try and "fail forward," if necessary boldly. Therefore, a Ph.D. study provides abundant creative opportunities for cultivating your unique research and curiosity style.

In addition to the opportunities to shape and be yourself, the thinker development opportunities, the leadership opportunities, the time-creation opportunities, and the service opportunities briefly discussed in Sects. 9.6, 9.5, 9.4, 9.3, and 9.2, respectively, a Ph.D. study also provides many miscellaneous personal/professional development opportunities, as summarized below.

9.7 Miscellaneous Opportunities

Strive not to be a success, but rather to be of value.

—Albert Einstein (1879–1955)

As miscellaneous takeaways of this chapter, the miscellaneous personal development opportunities of a Ph.D. study—but not limited to—are itemized below.

! Attention: Miscellaneous Opportunities of a Ph.D. Study

- *A Ph.D. is not enough.*
- *STEM is much broader than your Ph.D. project.*
- *STEM is much deeper than your Ph.D. project.*
- *Thinking about interdisciplinary, multidisciplinary, and transdisciplinary STEM.*
- *Thinking about the scholarship of application.*
- *There are always networking opportunities.*
- *There is always room for improvement.*

Next, we briefly discuss each itemized personal development opportunity.

9.7.1 A Ph.D. Is Not Enough

No technical skill is worth more than knowing how to select exciting research projects. Regrettably, this vital ability is almost never taught.

—Peter J. Feibelman

According to Peter J. Feibelman, who is the author of a thoughtful book entitled *A Ph.D. Is Not Enough! A Guide to Survival in Science* [10], a Ph.D. is not enough, especially when you target a bright research future with a successful academic or industrial career. Because a Ph.D. study often addresses a few research problems in a given narrow research field, earning a Ph.D. can make you know nearly nothing about the entire depth and breadth of developments in your research discipline. Therefore, a Ph.D. is not enough, which you shall seize as an immense opportunity to curiously, objectively, and persistently advance your intellectual capabilities and skills throughout your post-Ph.D. professional or research career.

9.7.2 STEM Is Much Broader Than Your Ph.D. Project

My brain is only a receiver. In the universe there is a core from which we obtain knowledge, strength and inspiration. I have not penetrated into the secrets of this core, but I know that it exist.

—Nikola Tesla (1856–1943)

Attributed to the nature of Ph.D. research, your Ph.D. research project—directed by your Ph.D. (co-)supervisor—will likely revolve narrowly on a limited scientific phenomenon, which is sought to be understood in greater depth. However, it is undeniable that STEM is much broader than your Ph.D. research project, whether it be complex STEM (e.g., *complex systems*), convergence STEM, or fundamental STEM. This is a gigantic opportunity for you to think in breadth on top of depth [11] so that you will likely succeed in a prospective academic/industrial career of impact.

9.7.3 STEM Is Much Deeper Than Your Ph.D. Project

To do successful research, you don't need to know everything, you just need to know one thing that isn't known.

—Arthur L. Schawlow (1921–1999)

Although *the sky is the limit* regarding Ph.D. research and academic career, there is always a limited scope in a research depth (i.e., an intellectual depth) typically achieved by most Ph.D. students. To this end, you shall be cognizant that science and engineering are much deeper than your Ph.D. research project. In doing so, you could seize the opportunity to challenge yourself and your intellectual development to pave the way to contributing an impactful depth in STEM.

9.7.4 Thinking About Interdisciplinary, Multidisciplinary, and Transdisciplinary STEM

Set a goal so big that you can't achieve it until you grow into the person who can.

—Zig Ziglar (1926–2012)

Because we are living in the era of *science and technology convergence* [7] (i.e., *the third revolution* [8]), you may target convergence research with meaningful impact on benefiting humanity, our planet, and the universe. To this end, you may often think of interdisciplinary, multidisciplinary, and transdisciplinary STEM—in connection with your Ph.D. research project—toward your discovery of impactful STEM convergence.

9.7.5 Thinking About the Scholarship of Application

It doesn't matter how beautiful your theory is, it doesn't matter how smart you are. If it doesn't agree with experiment, it's wrong.

—Richard P. Feynman (1918–1988)

Even though the *scholarship of application* [11] may not necessarily be the focus of your academic Ph.D. research project, you may also commit yourself to enthusiastically pursue vital scholarship of application—in relation to your Ph.D. research project of depth—provided that you aim for impactful works that would benefit humanity, our planet, and the universe.

9.7.6 There Are Always Networking Opportunities

Instead of better glasses, your network gives you better eyes.

—Ronald Burt (1949–)

Successful professional networking can create many opportunities that can inform and enhance not only your professional development but also your personal development. Accordingly, using the various twenty first-century technologies and the ubiquity of mobile Internet wisely, you need to not only seize several networking opportunities, which are always available, but also create crucial networking opportunities.

9.7.7 There Is Always Room for Improvement

Change is hard at first, messy in the middle, and gorgeous at the end.

—Robin Sharma (1964–)

Whether it be your writing, derivations (proofs), reasoning, thinking, paper introduction, literature survey, paper presentation, paper organization, paper discussion, paper methodology, paper simulations and computer experiments, and so forth, there is always room for improvement! To this end, if you often attempt to employ intellectual standards—such as *clarity, accuracy, relevance, logicalness, breadth, precision, significance, completeness, fairness, and depth* [12]—for evaluating each of your prospective work and draft paper, you will easily realize that each of your work and paper would need much more improvement in regard to each intellectual standard.

9.8 Concluding Remarks

> *Growth is the great separator between those who succeed and those who do not. When I see a person beginning to separate themselves from the pack, it's almost always due to personal growth.*
>
> *—John C. Maxwell (1947–)*

Intending to make you seize and create many personal development opportunities during your Ph.D. study, this chapter presented the personal development opportunities of a Ph.D. study. Specifically, we emphasized the service opportunities of a Ph.D. study, the time-creation opportunities of a Ph.D. study, the leadership opportunities of a Ph.D. study, the thinker development opportunities of a Ph.D. study, the Ph.D. study opportunities of shaping and being yourself, and the miscellaneous opportunities of a Ph.D. study.

Although we tried to discuss the most crucial themes concerning the personal development opportunities of a Ph.D. study, our discussion of limited scope—due to a space constraint—may not have covered the *full-blown depth and breadth* of issues intertwined with the discussed subjects. For detailed insights and realities, we encourage you to reflect this chapter's references and discuss them with your network of peers.

References

1. Wikipedia (2025) The Buddha. https://en.wikipedia.org/wiki/The_Buddha
2. Wikipedia (2025) Brian Tracy. https://en.wikipedia.org/wiki/Brian_Tracy
3. Wikipedia (2025) Warren Buffet. https://en.wikipedia.org/wiki/Warren_Buffet
4. Burchard B (2017) High performance habits: how extraordinary people become that way. Hay House, Carlsbad
5. Wikipedia (2025) Hyperfocus. https://en.wikipedia.org/wiki/Hyperfocus
6. Hill N (1937) Think and Grow Rich, Electronic edn. (2007)
7. Bainbridge WS, Roco MC (eds.) (2016) Handbook of science and technology convergence. Springer, Switzerland
8. MIT Washington Office (2011) The third revolution: the convergence of the life sciences. Physical Sciences, and Engineering. Washington DC
9. Guo PJ (2015) The Ph.D. grind: a Ph.D. Student Memoir. http://www.pgbovine.net/PhD-memoir.htm. Cited Jan. 2020
10. Feibelman PJ (2011) A PhD is not enough!: a guide to survival in science, revised edn. Basic Books, New York
11. Reis RM (1997) Tomorrow's Professor: preparing for academic careers in science and engineering. IEEE Press, New York
12. Paul RW, Elder L (2002) Critical thinking: tools for taking charge of your professional and personal life. Prentice Hall, Upper Saddle River

Index

T. M. Getu, *Pathways for Inspired Scientists and Engineers*,
https://doi.org/10.1007/978-3-032-06784-5

MIX
Papier aus verantwortungsvollen Quellen
Paper from responsible sources
FSC® C105338

FSC
www.fsc.org

If you have any concerns about our products,
you can contact us on
ProductSafety@springernature.com

In case Publisher is established outside the EU,
the EU authorized representative is:
Springer Nature Customer Service Center GmbH
Europaplatz 3, 69115 Heidelberg, Germany

Printed by Libri Plureos GmbH
in Hamburg, Germany